Palgrave Studies in Science, Knowledge and Policy

Many of the questions which concern us in our social, political and economic lives are questions of knowledge, whether they concern the extent and consequences of climate change, the efficacy of new drugs, the scope of surveillance technologies or the accreditation and performance of individuals and organizations. This is because what we know - how we acquire and apply knowledge of various kinds - shapes the ways in which problems are identified and understood; how laws, rules and norms are constructed and maintained, and which goods and services offered to whom. 'Who gets what, when, how', in Lasswell's phrase, depends very much on who knows what, when, how.

In our personal, professional and public lives, knowledge is a key resource. It matters in policy not only as a guide to decision making but because, in many circumstances, to be knowledgeable is to be powerful. Some kinds of knowledge are created and held by small numbers of specialists, while others are widely distributed and quickly shared. The credibility and authority of different kinds of knowledge varies over time and our means of developing and sharing knowledge are currently undergoing rapid changes as new digital technologies and social media platforms emerge. This book series is an interdisciplinary forum to explore these issues and more. In short, we are interested in the politics of knowledge. The series encompasses diverse topics, methods and disciplines and we welcome proposals for solo-authored, co-authored and edited books. Please contact the series editors, Kat Smith (Katherine.smith@ed.ac.uk) and Sotiria Grek (Sotiria.Grek@ed.ac.uk) to discuss your initial ideas and outline proposals. Kat Smith and Sortiria Grek are the Co-Directors of SKAPE (the Centre for Science, Knowledge and Policy) at the University of Edinburgh, UK. http://www.skape.ed.ac.uk

More information about this series at
http://www.palgrave.com/gp/series/14592

Marcos González Hernando

British Think Tanks After the 2008 Global Financial Crisis

palgrave
macmillan

Marcos González Hernando
Department of Sociology
University of Cambridge
Cambridge, UK

Palgrave Studies in Science, Knowledge and Policy
ISBN 978-3-030-20369-6 ISBN 978-3-030-20370-2 (eBook)
https://doi.org/10.1007/978-3-030-20370-2

This Palgrave Macmillan imprint is published by the registered company Springer Nature Switzerland AG
The registered company address is: Gewerbestrasse 11, 6330 Cham, Switzerland

To my family, old and new.

Foreword

Riders on the Storm: British Think Tanks and the Financial Crisis

Described a generation ago as merely 'second hand dealers of ideas,' today think tanks are often first in the frontline of contemporary policy debates. Think tanks provide 'explanations' for interpreting the causes and consequences of problems to help policymakers and publics understand the challenges they face. It is at the core of think tanks' 'business' to propose policy ideas and promote their recommendations. In order to do this, and be sustainable, they need policymakers, as well as society at large, to trust that they possess relevant knowledge and expertise which decision-makers can act upon. That is, think tanks must not only hold epistemic authority, but must be seen to wield it by 'truth-seeking'—through sound and rigorous analysis—rather than being motivated by what is economically, politically, or otherwise convenient. As such, the institutes that are the subject of this volume were thrown into the Financial Crisis—a storm they failed to predict, yet were positively energised by.

No matter their very different positions on the ideological and political spectrum, all four British think tanks that are the centre of analysis seek to be seen as 'sound' in their analysis and 'moderate' in their recommendations. As Marcos González Hernando ably recounts, their organisational careers in the crisis have been marked by successes with regards to gaining policy access and media acclaim, but their experiences in the policy debate are also littered with failures and false starts. He argues that the majority of previous studies on think tanks prioritise understanding their policy impact and thereby miss half the story; delving into the organisational 'black box' to scrutinise their internal functioning is just as fascinating. As a rule, he claims, in both ideological and material terms, think tanks are much more unstable and complex than they appear to outsiders.

The 2008 crisis caught many think tanks off-guard—as was the case with most political scientists and economists. Nevertheless, it was a 'fateful moment' providing a window of opportunity for the public, politicians, and policymakers to reflect profoundly upon the foundations of our socioeconomic order. As Marcos González Hernando argues, "this was not only an economic crisis but also an epistemic, political, and perhaps even a moral one." Given their *raison d'être*, think tanks were strongly compelled to offer explanations of the crisis and intervene in debates on the repercussions of the crisis. "Without an explanation, one cannot offer advice" González Hernando argues, and then outlines how the four performed predictably with regards to their 'public interventions' in the policy debate:

> […] a left-of-centre think tank would criticise the deregulation of the financial industry, and a sense of disappointment would follow after austerity became official policy. Conversely, a free-market think tank would be expected to chastise regulation and blame public officials rather than the private sector, greed, or the free market itself. An academic-technocratic organisation would most likely try to present itself as a neutral arbiter following evidence and arguing for its supremacy, and one supporting a party in government would devise a plan and—if seen to be animating actual policy—defend it through its implementation.

Notwithstanding their adherence to this 'script,' these think tanks were dramatically impacted by the crisis, especially in how they engaged with their publics, how they fashioned their analysis, and how they managed their internal affairs and in-house scholarly differences on policy issues. Think tanks are under pressure; this is the central theme of this volume.

In one sense this pressure can be seen in material terms. Worldwide, the think tank industry has not emerged unscathed from the Global Financial Crisis. Although there is little systematic available data, anecdotal accounts strongly suggest conditions of economic austerity have entailed lean times for think tanks. Over the past half century, the British think tank industry has grown, yet it seems the growth of existing funding sources and the development of new streams of donor support have proceeded at a slower pace. Even though British Government agencies and ministries are reliant on external expertise and have been a relatively consistent source of support for think tanks, the political field has become increasingly hermetic to external sources of advice and expertise. Besides, even at the best of times, funding agency preferences can be fickle. The UK's economic and social science research grant programmes increasingly impose guidelines on research activity. Wealthy individuals, corporate donors, charitable trusts, government departments, EU institutions, local government, and trade organisations have their own requisites, not only on the research agenda but also on how research output is delivered and disseminated. Resources are spread more thinly as donor expectations of policy impact increase. These circumstances have not only heightened competition among think tanks but also with other purveyors of policy analysis.

Concomitantly, the 'policy wonk' has become ubiquitous in more and more walks of life. Today there is more interchangeability between the 'think tank' and the 'academic' public intellectual. Universities have established their own institutes and 'policy labs' which sometimes rival think tanks. Many of the world's leading NGOs have developed a strong in-house capacity for policy research. Business associations, multinational corporations, banks, professional bodies, and trade unions are becoming better able to proselytise their policy perspectives with in-house research units or through more skilled use of social media.

Advances in telecommunications, as well as social networking, have fundamentally altered the environment in which think tanks operate. The twentieth century witnessed the rise of the think tank as an organisational form; might the twenty-first presage their decline? This is unlikely, Marcos González Hernando argues, due to the 'plasticity' of think tanks and how their organisational identities 'cross boundaries.' Through their oftentimes mobile and malleable workforce, think tanks are able to keep a presence in the media and NGO worlds while also engaging political and policy elites.

The four British think tanks in this study adapted constantly. They built sophisticated websites and professionalised their in-house communications offices. They jockeyed for position among different audiences via social media, press events, and closed meetings. And think tanks do retain some competitive advantages. The 'revolving door' benefits provided by think tank have not yet been undermined. Retiring politicians, diplomats, and civil servants will continue to seek out their post-retirement perch in a policy institute. Governments and the media will continue to seek out quality analysis and 'sound-bites' from think tank experts—at least for as long as think tanks position and project themselves as academically reliable.

Yet, popular disenchantment with the role of experts has ramped up societal questioning of think tanks. In our hyper-partisan world, the think tank industry often sharpens divisiveness rather than understanding in public debates. Marcos González Hernando suggests that some think tanks have helped undermine their own sources of authority. That is, societal unease and uncertainty over our capacity to describe the world (an epistemic crisis) has fostered a wariness over the trustworthiness of traditional sources of expert knowledge (a generalised crisis of expert authority).

Such scepticism of think tanks is, I would suggest, embodied in the growth of both *anti-politics* and *de-politicisation*. *Anti-politics*, on the one hand, is associated with a democratic malaise that takes expression in low voter turnout, electoral volatility, and the rise of protest votes and populist parties. *De-politicisation*, on the other hand, is a tactic usually employed by governments to deflate contentious policy issues and make them seem technical and neutral.

De-politicisation through preference-shaping and agenda-setting speaks to the establishment of a 'dominant rationality' and non-decision-making dynamics that systematically delete certain problems or issues from public debate and policy consideration. Expertise—such as that commissioned from think tanks—is deployed to help entrench a certain way of 'seeing' and defining problems, the elaboration of ideological infrastructure and public values of such problem definition, and the development of models and methodologies to 'manage' problems. Accounts of the impacts of neoliberal and technocratic think tanks are often detailing this kind of *de-politicisation* tactic.

Scientisation is a related tactic of *de-politicisation* but one connected much more closely with 'experts.' Due to rapid technological change and scientific advances, most fields of public policy have become highly complex, requiring regular input and monitoring by trained professionals and scientific advisors. Reliance on expert consultation, evidence construction, and technocratic deliberation in local, national, and global governance, institutes 'knowledge' organisations as governance institutions in their own right. The arcane interests, the professional communication codes, and technocratic character of many think tanks puts them at a distance from wider society and popular concerns. Traditionally, think tanks have faced power rather than the public.

Rather than arguing that this is an inherently apolitical dynamic, in the sense of completely foreclosing dissent, epistemic power is in constant contest. There are challenges to dominant knowledge groups from norm-based groups and networks in civil society providing alternative visions of policy and engaged in the repoliticisation of neutral economic theory or policy orthodoxy. The competition of ideas is a never-ending struggle between different 'worldviews' and 'regimes of truth.' Rather than think tanks 'speaking truth to power,' with their primary focus upon government, in an era of anti-politics and austerity, think tanks are potentially decentred towards communities and citizenries in the construction of policy analysis. That is, they may intervene in public debates in a way that facilitates awareness and dialogue among plural sets of stakeholders to social and economic problems. Although we arrive at this point from different directions, this gradual repositioning of think tanks towards many different publics is one of the more interesting analytical insights from Marcos González Hernando.

To end on an optimistic note, think tanks are not irretrievably depoliticised as 'scientised' tools exploited by governments and decision-making elites. Given their propinquity to valorise and prioritise questioning in public deliberation and critical analysis in policy debates, most think tanks remain reliable instruments for policy reflection and socioeconomic analysis. Through their frequent policy positioning in order to capture the political imagination of different audiences with different interests, think tanks resonate with a range of intellectual and institutional transformations.

Budapest, Hungary

Diane Stone
Dean, School of Public Policy, Central European University

Centenary Professor of Governance and Policy Analysis, University of Canberra

Vice President of the International Public Policy Association

Acknowledgements

I would like to start by thanking my interviewees, who so kindly offered their time and insights. Meeting them taught me much about their fascinating organisations and even about myself.

I also wish to thank the scholars working on think tanks, intellectuals, political sociology, and public policy who guided me through this process. My supervisor, Patrick Baert, for his support and clarity; Diane Stone, for her penetrating ideas, goodwill, and, of course, her generous preface; Gil Eyal and Hazem Kandil, for their comments on the doctoral dissertation on which this book is based; as well as Filipe Carreira da Silva, Gisèle Sapiro, Darin Weinberg, Jeff Miley, Vineeta Sinha, Michèle Lamont, Simon Susen, John Holmwood, Christina Boswell, Fiorenza Gamba, Katarzyna Jezierska, Stella Ladi, Vikki Bell, Don Slater, Maite de Cea, Hartwig Pautz, Dieter Plehwe, Jordan Tchilingirian, Karin Fischer, Erin Zimmerman, Kate Williams, and Julien Landry. Their writings and observations improved my ideas tremendously.

I should also thank my fellow friends and Ph.D. students from this beleaguered generation, in Cambridge and elsewhere. In no particular order, they are Kusha, Carla, Magdalena, Naim, Tiago, Eric, Eliran,

Mahvish, Tobias, Weeda, Olga, Robert, Phillip, Macarena, Ozan, Yesim, Christina, Rin, Belén, Diane, José, Tomás, Ignacio, Christina, Torsten, Linzhi, Nurjk, Ellen, Liz, Izabela, Blair, Martina, Marcus, Amín, and Amine. Each of them, in their own way, shaped my thinking and made me realise the extent of my ignorance. And though I did not have the pleasure of knowing him closely, I would also like to pay tribute to Giulio Regeni, for his memory and example has inspired the generation of social scientists of which I consider myself a part.

I must also mention the institutions that assisted or supported me throughout these years: FEPS and TASC, especially Shana, Rob, Sidney, Tyler, Kirsty, John, and Sylvia; Neal, Gerry, Frances, and Remco from Compass; Christian from *Distinktion*; Gabriel, Pablo, and Miguel from Ballotage; Enrique and Andrea from OnThinkTanks; Rosemary from the LSE Review of Books; those I was fortunate enough to meet at l'EHESS in Paris and, of course, at the University of Cambridge and Fitzwilliam College.

These acknowledgements could not miss Darja Irdam, Katie Gaddini, Peter Walsh, and Jordan Tchilingirian (again!) for their friendship and their comments on earlier versions of this book; Chloe Clifford-Astbury and Deborah Huyton, who were of great help with transcriptions and coding; and Beth Farrow, Poppy Hull, and Tamsine O'Riordan from Palgrave, as well as Katherine Smith and the scholars at SKAPE, University of Edinburgh, whose input has made this book infinitely better.

I also want to thank my dear friends Carmen, Fabien, James, Murari, Thiago, Marita, Elina, Sam, Ogi, Fabio, Jerome, Dimuthu, Bojana, Nat, Andrei, Alex, Coral, Paula, Roberto, Franko, Eduardo, Rodrigo, Claudio, Pablo, Liam, Franco, Iván, Adriana, Loreto, Manuela, Marcela, Alice, Juliane, and Makis.

But most of all I thank my family. My parents Marco and Marcela, and sisters, Daniela and María Francisca, a constant source of inspiration, admiration, love, and support; and Irina, mi compañera. Without them this book would have been impossible, and I would be much poorer.

Marcos González Hernando

Contents

About the Author

Marcos González Hernando is Affiliated Researcher at the University of Cambridge, Senior Researcher at Think Tank for Action on Social Change (FEPS-TASC), and Managing Editor at *Distinktion: Journal of Social Theory*.

Abbreviations

AFPX	American Friends of Policy Exchange (PX)
ASI	Adam Smith Institute (UK)
BIS	Department for Business, Innovation and Skills (UK)
BoE	Bank of England (UK)
CASE	Centre for the Analysis of Social Exclusion (UK)
CEP	Centre for Economic Performance (UK)
CESifo	Center for Economic Studies and Ifo Institute Group (Germany)
CFM	Centre for Macroeconomics (UK)
CFR	Council of Foreign Relations (USA)
CPS	Centre for Policy Studies (UK)
CSJ	Centre for Social Justice (UK)
DEFRA	Department for Environment, Food and Rural Affairs (UK)
DfID	Department for International Development (UK)
DWP	Department for Work and Pensions (UK)
ECB	European Central Bank
ESRC	Economic and Social Research Council (UK)
ESRI	Economic and Social Research Institute (Ireland)
EUROFRAME	European Forecasting Research Association for the Macroeconomy
FES	Friedrich-Ebert Stiftung (Germany)

FSA	Financial Services Authority (UK)
FT	*Financial Times*
GND	Green New Deal (NEF)
HECSU	Higher Education Careers Service Unit (UK)
HPI	Happy Planet Index (NEF)
ICB	Independent Commission on Banking (UK)
IEA	Institute of Economic Affairs (UK)
IFS	Institute for Fiscal Studies (UK)
INET	Institute for New Economic Thinking (USA)
IZA	Forschungsinstitut zur Zukunft der Arbeit (Germany)
IPPR	Institute for Public Policy Research (UK)
JRF	Joseph Rowntree Foundation (UK)
LLAKES	Centre for Learning and Life Chances in Knowledge Economies and Societies (UK)
LSX	London Stock Exchange (UK)
MEP	Member of European Parliament
MP	Member of Parliament (UK)
MPC	Bank of England's Monetary Policy Committee (UK)
NAO	National Audit Office (UK)
NBER	National Economic Research Bureau (USA)
NCPA	National Center for Policy Analysis (USA)
NEF	New Economics Foundation (UK)
NEON	New Economics Research Network (NEF)
NESTA	National Endowment for Science, Technology and the Arts (UK)
NHS	National Health Service (UK)
NIER	National Institute Economic Review (NIESR)
NIESR	National Institute of Economic and Social Research (UK)
NiGEM	National Institute Global Econometric Model (NIESR)
OBR	Office for Budget Responsibility (UK)
ODI	Overseas Development Institute (UK)
PRIME	Policy Research in Macroeconomics (UK)
PSI	Policy Studies Institute (UK)
PX	Policy Exchange (UK)
QE	Quantitative Easing
RSA	Royal Society of Arts (UK)
SPD	Sozialdemokratische Partei Deutschlands (Germany)

SROI	Social Return on Investment (NEF)
SSRC	Social Science Research Council (UK)
TNG	The Next Generation (ASI)
TTIP	Transatlantic Trade and Investment Partnership
TUC	Trades Union Congress (UK)
WIIW	Wiener Institut für Internationale Wirtschaftsvergleiche (Austria)
WZB	Wissenschaftszentrum Berlin für Sozialforschung (Germany)

List of Figures

List of Tables

1

Thinking Under Pressure: Think Tanks and Policy Advice After 2008

> An explanation was needed, one was found; one can always be found; hypotheses are the commonest of raw materials.
>
> Henri Poincaré, on Lorentz's theory of aether. (Bourdieu 1988: 159)

On Monday, 15 September 2008, the investment bank Lehman Brothers filed for bankruptcy after the US government decided not to provide it with emergency liquidity. It had already done so in the preceding months for three other large financial institutions—Bearn Stearns, Fannie Mae, and Freddie Mac—and enough was enough. That week, stock markets across the world went into tailspin and governments rushed to scrap together bailout plans of bewildering proportions. The world economy entered its worst recession since 1929. Mainstream economists—who dominated thinking in policymaking, finance, and the social sciences—had told us this was just not possible. A few years before, Nobel Memorial Prize in Economic Sciences winner Robert Lucas Jr. had opened his American Economic Association's presidential address with the following words:

M. González Hernando, *British Think Tanks After the 2008 Global Financial Crisis*, Palgrave Studies in Science, Knowledge and Policy, https://doi.org/10.1007/978-3-030-20370-2_1

> Macroeconomics was born as a distinct field in the 1940s, as a part of the intellectual response to the Great Depression. The term then referred to the body of knowledge and expertise that we hoped would prevent the recurrence of that economic disaster. My thesis in this lecture is that macroeconomics in this original sense has succeeded: Its central problem of depression prevention has been solved, for all practical purposes, and has in fact been solved for many decades. (Lucas 2003)

Whenever there is a crisis, there are demands for an explanation. The 2008 global financial crash gave rise to countless diagnoses of what went wrong, which had to readdress the foundations of the economic order of modern capitalist societies. Suspicions on the sustainability of the financial industry and free markets took centre-stage in a way they had not since at least the 1990s 'end of history' era. While the immediate trigger of the near-collapse of the banking system was commonly traced to the expansion and bursting of the US subprime mortgage market, ascertaining its ultimate causes and consequences became the subject of much controversy. As such, the events of 2008 amounted not only to a market crash, but to a crisis of self-understanding, and could be considered a sudden trauma that demanded interpretation (Eyerman 2011).

Given these circumstances, both new and established political actors and policy experts were impelled to make their case in the public arena. Indeed, it became nigh unavoidable for them to do so, as explanations were widely sought by the public and offered by competitors. This unsettled environment could, lest we forget, create the conditions for momentous political, economic, and societal change. Disseminating one's interpretation of the situation became a critical goal, as it could determine the decline or rise to prominence of one's ideas and, not uncommonly, of one's career (Campbell 2002).

In this milieu, expertise had a central part to play (Brooks 2012). The precipitous collapse of major financial institutions, and its drastic effects on national economies, spurred a plethora of technical and moral explanations of what happened, how expectable it was, and what should be done in its wake (e.g., de Goede 2009; Sinclair 2010; Lo 2011; Rohloff and Wright 2010; Thompson 2012). Even so, questioning traditional experts also became commonplace, as most failed to foresee (and some even declared impossible) what they were now advising on (Engelen et al. 2011). Years later, Bank of England's Chief Economist Andrew Haldane said that the financial crash was economics' 'Michael Fish moment,' in reference to a BBC weather forecaster who in 1987 dismissed warnings of an impending hurricane just hours before South East England suffered the worst storms in three centuries (BBC 2017).

In time, the initial confusion gave way to more theoretical discussions on the dangers of complex financial instruments, the sustainability of credit-driven growth, the role states and regulators should play in steering (or setting free) the economy, and on whether to prioritise tackling public deficits or stimulating demand. The crisis certainly energised debates on financial, fiscal, and monetary policy, and even prompted the comeback of the ideas of classical economists such as Smith, Marx, Keynes, and Hayek (Solomon 2010). In sum, the events of 2008 brought about a growing public interest in economics, both as a subject and as a profession (Fourcade 2009; Gills 2010). Because of the above, debates on economic policy after the crisis often went beyond economics narrowly understood, especially once it became uncontroversial that matters that had hitherto been the almost exclusive remit of economists could not be left solely to experts with bounded rationality (Bryan et al. 2012). Bluntly put, the crisis marked the beginning of the end of the 'centrist,' 'technocratic' divorce between politics and economics, and of the subordination of the former to the latter.

Around the same period, the most important online social media platforms were in their early days. What are nowadays household names such as YouTube, Twitter, and Facebook were in 2008 only a few years old and growing at full tilt. New forms of producing, consuming, and disseminating knowledge became ever more ubiquitous, which prompted political actors to supply increasingly more accessible and well-targeted content through these channels (Brooks 2012). These new means of communication offered abundant opportunities for individuals and organisations to broadcast their views to a more attentive public. Yet, that public was also more sceptical. The urgent need to make sense of the economy, just after the dominant view of economics had been found ill-prepared to do so, presented threats and opportunities to policy experts and politicians alike (Aupers 2012; Rantanen 2012). In this context, the mission of think tanks—institutions at some middle point between being academic bodies, media commentators, political actors, and lobbying organisations—was to inform the political debate and influence policy. These organisations and their public interventions are the subject of this book, through which I seek to contribute to the sociology of knowledge and expertise.

Why Think Tanks?

This book derives from a doctoral dissertation in sociology, which I undertook because I was curious about how the ideas of those whose job is to produce them change over time, especially after the foundations of their legitimacy have been shaken. My interest in the 2008 crisis was obvious enough, and the centrality of the UK in the global financial system made it a privileged backdrop. I decided to focus on think tanks mainly because of two reasons.

Firstly, because their very business is to propose and promote ideas that have a bearing on politics and policy. For these to find a wide hearing, think tanks need to be seen as credible 'experts' by their relevant publics or, in other words, to have some measure of expert (or epistemic) authority. Throughout this book, by expert authority I mean the capacity of an actor to be trusted to possess and produce knowledge which others can responsibly refer to and base their decisions upon (Pierson 1994; Herbst 2003). Traditionally, yet not always, expert authority is linked to being perceived as having some degree of cognitive autonomy, meaning an ideal condition for knowledge production in which pronouncements on a subject are thought to be based on reasoned argument alone. That is, actors are thought to have cognitive autonomy if their statements are perceived to be animated mainly by 'truth-seeking' rather than by what is economically, politically, or otherwise expedient. On that line, while some readers might be suspicious of readily applying the word 'expert' to think tanks, I do so because their objective is precisely to be seen as such. Thus understood, 'expertise' is, before anything else, a type of social relationship (Eyal and Pok 2011).

If ever achieved, the expert authority of a think tank is an unstable accomplishment that depends on its reputation across many audiences that might have very different world views, and which can change over time. That is the second reason behind my interest in these organisations. Their murky character, hovering over the edges of advocacy, academia, economic interests, and politics (Medvetz 2012a), renders think tanks a privileged index of their environment. Their 'boundary-crossing' makes think tanks noteworthy artefacts of modern politics, as their relevance depends on a complex bundle of capabilities and resources. Links to academics, journalists, charitable and corporate donors, third sector organisations, civil servants, and politicians can be critical assets for think tanks, shaping their image, fundraising capacities, and research outputs (McNutt and Marchildon 2009). In other words, think-tankers, whose goal is to be seen as politically and intellectually relevant and attract supporters, must learn to garner, use, and dispose of diverse types of resources in a rapidly shifting environment in which the 'worth' of these resources is never settled. They need to play several games at once, whose rules might change mid-game, and with the caveat that winning in some might mean losing out in others.

In more practical terms, this book focuses on how four theoretically sampled British think tanks sought to make sense of the economic crisis and convince others of their account of events. These are the New Economics Foundation, the Adam Smith Institute, the National Institute for Economic Affairs, and Policy Exchange. By analysing their publications, their annual accounts, their organisational structure, and their presence in the media, and aided by interviews with current and former members of staff, I trace the process by which they reacted to unfolding events and were transformed by the crisis of a decade ago. I report how think tanks intervened on public debates on economics and finance, the extent to which these public interventions betrayed changes in their organisations, and how these changes reflect their broader environment. In the end, I argue that in 2008 an epistemic crisis (uncertainty over our capacity to describe the world) was associated with a generalised crisis of expert authority (a wariness over the trustworthiness of traditional sources of expert knowledge). These developments had at least two major effects on think tanks: it made it more difficult for them to reach beyond their established constituencies—those who do not broadly share their views on politics and economics—and reduced the penalty for ignoring the consensus from mainstream sources of expertise (e.g., academic economics).

Since their inception, think tanks have been beset by doubts over their independence. While nearly all declare themselves to be (at least formally) politically and intellectually autonomous, questions over their disinterestedness are not uncommon. This is a central issue of contention, as often a precondition for garnering expert authority is to be perceived as having some level of cognitive autonomy. That is why much of the scholarship on think tanks delves in length over the extent of their professed independence from the powers that be (see Stone 1996; Abelson 2002). Though it is not my wish to claim think tanks are either mouthpieces for vested interests or completely self-determining, I posit that tracing how they change over time can cast a light on what sort of pressures they experience. Leaving aside for a moment the question over their independence, focusing on the reaction of think tanks to a crisis of expert authority can help us understand their wider environment. After all, no organisation is completely devoid of external influences. Given the above, this research has as background the sociology of knowledge, especially in relation to intellectual change, the framing of crises, and the capacity of policy actors to procure expert authority for themselves while undermining the claims to expertise of others (see Beck and Wehling 2012; Davies and McGoey 2012; McGoey 2012).

Research on think tanks, and in particular those in Britain, has been until recently relatively scarce and has been mostly preoccupied with whether and how they shape public policy (e.g., James 1993; Tesseyman 1999). Apart from Pautz (2012a, b, 2016), Bentham (2006), and Denham and Stone (2004), and a small but growing number of others, most academics have concentrated on their rise to prominence during the 1970–1980s in the context of the end of the post-war consensus, or on their effects on specific policy areas—e.g., cultural policy, education, healthcare (Schlesinger 2009; Ball and Exley 2010; Kay et al. 2013). Few study in much depth how they change over time, preferring to concentrate on their links to party elites and their policy impact—though see McLennan (2004). Similarly, scholarly work directly focused on think tanks in relation to the 2008 economic crisis is only now emerging—though see issue 37:2 of *Policy & Society* (González Hernando et al. 2018).

Although, unavoidably, this book has some bearing in economics, its focus is on the production of knowledge about the economy rather than on explaining the 2008 crisis or evaluating the merit of the policies designed to address it. Instead, it is a second-order interpretation, an 'observation of observers.' In that ambit, there is a robust literature on the intellectual consequences of the 2008 crisis. Much of it focuses on economists, policymakers, and the persistence of neoliberal policies and ideas even after what many view as a challenge to their legitimacy (e.g., Gamble 2009; Lawson 2009; Crouch 2011; Schmidt and Thatcher 2013; Walby 2016; Tooze 2019). Another branch of the scholarship looks at how the 2008 crisis was framed by policy actors such as central banks, the media, corporations, politicians, and governments (e.g., Boin et al. 2009; t'Hart and Tindall 2009; Abolafia 2010; Sandvoss 2010; Lischinsky 2011; Banet-Weiser 2012; Berry 2016; Wren-Lewis 2018). This book touches on some of the issues that have troubled these authors, but rather than focusing on the effects of the crisis on public discourse or the impact of think tanks in shaping it, it inverts the centre of attention, zeroing in on think tanks themselves. In other words, this research takes organisational instability as its starting point, and rather than measuring the extent of think tanks' successes, it traces their actual work over time, thus providing an account littered with failures and false starts.

Analysing how think tanks weathered the crisis, both in intellectual and institutional terms, will contribute to our understanding of how organisations oriented towards the policy debate are affected by major external events. In practice, this means focusing on their public interventions, meaning any communicative act by which they seek to draw attention to their ideas—talks, policy reports, blogs, media appearances, parliamentary hearings, tweets, etc. Throughout this book, I show how think tanks changed the way they engage with their audiences, as wider transformations in the conditions that make their public interventions possible were underway, namely: shifting sources of funding; the rise of social media; a siloed media environment; a vague but growing mistrust of expert authority; and a political field that, while relying on expert discourse, became ever more hermetic to outside expertise.

To achieve the above, I employ conceptual tools drawn mainly from three sources: Thomas Medvetz's (2012a) Bourdieusian model of think tanks as organisations at the boundaries of social fields; the sociology of intellectuals and their interventions; and Neoinstitutionalist theories of organisations and policy change. Readers who would like to peruse through a more in-depth review of this literature—and my own theoretical and methodological contribution to it, as I see it—can find it in Chapter 2. In what remains of this introduction, I provide a brief exposition of the scholarship on British think tanks, expand on the logic behind this book, provide an account of how I understand think tanks as a research object, and introduce the four case-studies.

Think Tanks in Britain

Academic interest in think tanks, like the phenomenon itself, is relatively recent and began in the USA. Most of the early literature was structured around the divide between elite theorists and pluralists. The former, drawing from Mills (1956; Domhoff 1967) liken think tanks to lobbyists and pressure groups, their mission being to masquerade interests as research, while the latter believe them to be only one actor among many competing for attention in a crowded and pluralistic public debate (Abelson 2002). Both approaches have garnered criticism. Critiques of elite theories focus on their mechanistic description of think tanks and their lack of nuance when studying institutions with different degrees of closeness to power. Meanwhile, objections to pluralists concentrate on their neglect of power relations where they exist and their overuse of the accounts think tanks give of themselves, assuming too hastily their own claims to independence. Most researchers nowadays believe this discussion to be outdated, preferring to draw insights from both currents (Abelson 2012; Medvetz 2012a).

In Britain, the academic debate on the topic started in earnest with Richard Cockett's (1995) *Thinking the Unthinkable*, which examined the influence of new-right think tanks on the rise of Neoliberalism and on Thatcher's premiership. Cockett studies the historical conditions that allowed what was in the 1930s a disparate and relatively marginal group of thinkers to gradually gain intellectual legitimacy and political clout, among them Friedrich von Hayek. By following new-right think tanks and their members, Cockett gives a compelling account of how they spearheaded the rise of free-market liberalism and the undermining of the post-war consensus (see Muller 1996).

Cockett's book was followed by Andrew Denham and Mark Garnett's (1998) *British Think Tanks and the Climate of Opinion*, perhaps this book's most comparable precursor. Denham and Garnett provide a broad historical overview of the emergence of British think tanks, structured around five case-studies and divided in three waves. The first of these waves emerged in the interwar period and includes Political and Economic Planning (now Policy Studies Institute, PSI, est. 1931) and the National Institute of Economic and Social Research (NIESR, est. 1938). These *avant la lettre* think tanks were mostly research-oriented organisations that sought to directly connect the positivistic social sciences of their time with government in an 'enlightenment' model of policy influence. A second wave, the focus of Cockett's book, is more polemical and overtly political and includes the Institute of Economic Affairs (IEA, est. 1955), the Centre for Policy Studies (CPS, est. 1974), and the Adam Smith Institute (ASI, est. 1977). A third, more numerous and diverse wave emerges around the end of the Cold War, comprising left-of-centre responses to the second wave (e.g., the Institute for Public Policy Research, IPPR, est. 1988; Demos, est.1993), specialist think tanks (European Policy Forum, EPF, est. 1992), and offshoots from earlier institutions (Politeia, est. 1995, founded by former IEA staff).

Denham and Garnett take issue with the expression 'climate of opinion,' by which Cockett refers to the ideas that are dominant in the policy debate, and which he borrows from IEA's mission statement—written by Hayek himself. Denham and Garnett's main contention is that this idiom risks overstating the influence of new-right think tanks and taking their word at face value. This 'climate' might not be that "of the great outdoors but of a sedulously air-conditioned penthouse" (Guinness, in Denham and Garnett 1998: 200). Instead, they argue that it is the demands of policymakers for external legitimation rather than the weight of think tanks' ideas what grants them a semblance of influence. Politicians benefit from portraying sympathetic think tanks as authoritative, independent, and in tune with the 'climate of opinion,' as they can provide the impression that outside experts back their policies. Meanwhile, media outlets profit from giving exposure to think tanks as suppliers of off-the-shelf opinion and research to enforce their own biases or seek journalistic balance. Furthermore, think tanks themselves have a vested interest in exaggerating their importance and depicting public debate as a 'battle of ideas' (see also Krastev 2001). These reasons made Denham and Garnett wary of the risks these institutions could represent for a pluralistic democratic debate.

Further grounds for this wariness towards viewing think tanks in purely intellectual terms are expressed by Diane Stone (2007) in her article *Recycling bins, garbage cans or think tanks?* Stone debunks what she sees as three myths surrounding these organisations. The first of these is that they 'bridge' different domains (e.g., science and policy), as declaring think tanks to be autonomous of either part of what they connect mystifies their purpose and overstates their autonomy from either. Secondly, it is not self-evident that think tanks serve the public interest, as they often have close connections with interest groups and since many seek to influence and play to the biases of the media or a narrow elite rather than to inform the public (see Jacques et al. 2008). The third myth is that think tanks 'think' in the first place. Many, according to Stone, are re-packagers of previous research (recycling bins), a reservoir of policy solutions to be pushed whenever a relevant problem appears (garbage cans), or bodies that seek to grant socio-scientific validation to previously held views. Agreeing with these admonitions, towards the end of this book I argue that the capacity of think tanks to moderate the relationship between knowledge production and politics makes them privileged spaces for the curation and cultivation of politically fit expert knowledge.

Underlying these discussions, the figure of Antonio Gramsci looms large—a theorist deeply indebted to Machiavelli who thought of intellectuals in mostly antagonistic terms. Perhaps this is not surprising; we are speaking of 'tanks' after all. One of the first scholar to employ Gramscian ideas to the study of British think tanks was Rhadika Desai

(1994). She used concepts such as 'hegemony' and 'organic intellectuals' to explain how new-right think tanks contributed to the ascent of monetarism and the political shifts of the 1980s. After a historical account of the tensions underpinning British capitalism, Desai claims a moment of organic crisis opened the space for ideas hitherto marginal in economics and beyond what was then considered politically viable. The task of free-market think tanks was, according to her, to coordinate an intellectual attack on the Keynesian consensus by producing an alternative vision that could gain traction within economics departments, the Conservative Party, and beyond. Ultimately, this was an attempt by 'organic intellectuals' to convince and recruit 'second-hand dealers in ideas'—journalists, editors, commentators, educators—to conquer the 'common sense,' and with it, 'traditional intellectuals.'[1] This shift was achieved in no small measure through the undermining of any comprehensive opposition: those making the case for free markets needed not convert everyone, but merely to convince as many as possible that there is no feasible alternative.

More recently, Hartwig Pautz (2012a) employed Gramscian ideas to study the history of think tanks associated with left-of-centre parties in Germany and the UK around the end of the Cold War. He claims that, as the centre-left experienced an identity crisis, new think tanks flourished around the German SPD (FES, WZB) and the British Labour Party (IPPR, Demos), which veered these parties' views on the relationship between the state and the market. Like Desai, Pautz's main contention is that think tanks are most effective when the core tenets of an ideological position are under attack. Dieter Plehwe and Karin Fischer (Plehwe 2010; Fischer and Plehwe 2013) make another contribution to this line of thinking by analysing how think tanks partake in international networks. According to them, new-right think tanks have

[1]For Gramsci, traditional intellectuals are those who claim to seek truth, while organic intellectuals are those who represent the interests of a particular social position or class (Gramsci 1999[1971]). Also, by 'common sense' (from the Italian *senso comune*) Gramsci means the disparate set of ideas about the world that are held widely yet vaguely within a community, without the connotations of reasonableness and even-handedness that are present in its English counterpart (see Crehan 2016).

organised themselves in 'transnational discourse coalitions,' making ever more potent the argument that there is no alternative to neoliberalism.

Nevertheless, although Gramscian accounts of think tanks provide important background, this book is, in comparison to most of them, less ambitious and less directly concerned with the political to and fro. I argue that while thinking of think tanks only in cognitive terms can neglect the power relations in which they are embedded, doing so only in political terms can dampen our sensitivity to intellectual change. Moreover, these approaches differ from this research in that, by necessity, they mostly investigate think tanks obliquely, through their political effects rather than concentrating on what they actually do. And although Gramsci himself had a developed concept of intellectual change (see Crehan 2011, 2016), most Gramscian accounts of think tanks have tended to understand them as warring factions with little time for, or interest in, questioning their own views. To avoid the risk of having too stiff an image of think-tankers, I remain agnostic towards the possibility of them changing their minds or doing something one would not expect of them. The opposite would mean treating them as stooges from the get-go.

Be that as it may, the 2008 financial crash provides an excellent vantage point to examine how those tasked with informing the policy debate react to a crisis of the economy, politics, and expertise. Think tanks could certainly act as fortresses for intellectual 'foot-soldiers,' as many would anticipate, or adapt their thinking and strategy to a new era. Throughout the empirical sections of this book, I detect whether change occurs through what I call the *hysteresis* hypothesis: simply put, contrasting what they actually did with what an informed observer would predict (see Chapter 2). In that line, the next section sets forth some of the assumptions that underpin the design and structure of this research, as well as its scope—what this book is not, and what it seeks to be.

The Rationale Behind This Book

In general, the literature on think tanks could be said to be built over a duality inherent in its object of study—the intellectual and the institutional, the ideational and the prosaic. From these distinctions, one could understand these organisations as primarily guided by either their own self-preservation—in both financial and political terms—or by their professed intellectual commitments. Two ways of structuring a research project thus arise. A first would be to ascertain how a think tank's organisational characteristics and political alliances impact their behaviour. In other words, this would mean tracing how institutional pressures—most obviously funding and networks—shape a think tank's output. A second option would be to focus on the ideational and narrative aspects of their work; how they describe society and the economy, and the internal logic of this description.

Both these options are worth pursuing, yet insufficient on their own. Concentrating on only one aspect of think tanks—interests or ideas—risk reiterating either the crude materialism of some elite theorists, or the neglect of power relations of pluralists. Furthermore, think tanks are, both in their thinking and in their practice, as a rule more unstable and complex than they appear to outsiders. For those reasons, accounting only for interests or ideas—under the assumption that either is think tanks' ultimate driver—risks restricting our perceptiveness to change.

As stated earlier, the aim of this book is, neither, to measure the effectiveness of think tanks in affecting policy or the 'climate of opinion.' This is also because such a project would be methodologically tricky, on account of the difficulties involved in detecting their impact, where to seek it, and even how to define it. After all, most think tanks have, if any, only a nebulous imprint in public opinion and policy, which hinders efforts to measure where they were prominent and to what extent. Whenever the policy influence of think tanks is discussed, I rely mainly on think-tankers' own assessment—describing how their perceived successes or failures altered their course of action.

Instead, this book is structured around three questions: *what were the public interventions of these four think tanks on economic policy between and 2007 and 2013? What do these reveal of their intellectual and institutional transformations? and what do these transformations, in turn, say about their environment?*

By choosing this focus, I stress the importance of change (or lack thereof) as measurable through think tanks' output. Centring on their public interventions also avoids reifying them or delving into their motivations (as if they were thinking entities themselves), concentrating instead on what is 'uttered' in their name by actual individuals. Further, this focus facilitates detecting change both at the level of the substantive content of their work (which implies intellectual repositioning) and its format (which can be evidence of shifts in their target audience, their media strategy, and even their funding).

Six presuppositions are implicit in this research strategy. The first is that the financial crisis was an exogenous event. Think tanks did not produce it, at least directly. Indeed, it caught most by surprise, even if critics might reasonably claim it amounted to the unintended consequence of policies supported by some of them. Further, even if members of a think tank had foreseen its occurrence, they could do little to prevent it. The 2008 crisis was an issue think tanks had to react to.

Secondly, think tanks—especially those attentive to economic affairs—are strongly compelled to offer an explanation of the financial crash and produce work on at least some of its aspects and repercussions. The crisis was a momentous event, impossible to be ignored. What is more, in its wake it becomes almost compulsory for every policy expert to have a 'narrative' at hand. If a think tank is silent about the causes of the crisis, or on what could be done to remedy it, it risks trailing behind its competition. Without an explanation, one cannot offer advice.

The third assumption is that the financial crisis of 2008 was a 'fateful moment' that cannot be explained solely by appealing to particularistic or narrowly disciplinary reasoning. It was an event that made it more probable for actors to reflect upon the basic tenets of our economic and social order. Hence, it allowed for the emergence of public controversies—e.g., macroprudential regulation—that were hitherto the almost exclusive remit of specialists. With Boltanski and Thévenot (2006), one could claim the crisis urged actors to lay bare the fundamental justification of their everyday behaviour. This is especially true for the finance world and their political supporters. Expectably, many agents came forth to claim our knowledge of the economy to be either sufficient or not to explain what happened, to stress or downplay the historical significance of events, but the very emergence of these fundamental questions carries far-reaching implications. After all, this was not only an economic crisis but also an epistemic, political, and even a moral one (Morin 1976).

On this point, although a clear-cut distinction between 'crisis' and 'normality' can be problematic, instances such as the one discussed here are difficult to disentangle from discourses that take 'crisis' as their rhetorical basis. Notwithstanding some compelling arguments for challenging the concept of crisis (Roitman 2013), my more humble intention here is to explore how narratives about the crisis are formed and mobilised rather than to supersede them. Furthermore, for think tanks, the choice of which policy issue to concentrate on can be in itself telling. Said decision may imply that some factors are deemed more relevant than others, but also that more support is available to explore some of its aspects. This is why I do not employ a narrow or prescriptive definition of 'economic crisis.'

The fourth underlying premise is that political crises tend to have a trajectory (see Boin et al. 2009). Generally, a moment of eruption is followed by another of heightened uncertainty, which then opens a window of likely political change and, depending on the events that follow, leads to a 'condensation' of positions around the issue. This implies that timing is of the essence. Something 'said' in one moment may have dramatically different consequences depending on when it is uttered. It is very different to claim that the global financial system is unsustainable in 2006, 2009, 2013, or 2019. Furthermore, the trajectory of the global economic crisis varies across countries, and similar events can be construed as critical in one instance and not in another (Brändström and Kuipers 2003). One of the by-products of this book is therefore a history of the evolution of how the 2008 crisis was understood in different circles of British economic and social policymaking.

The fifth assumption behind this book is the double character, intellectual and institutional, of both think tanks and their experience of the crisis. Think tanks have an intellectual facet visible in both the tools they employ—e.g., academic references—and constraints they face—e.g., internal coherence, presumed shifts in their audiences. By their institutional facet I refer to, for instance, their funding, its volume and sources, their networks, their staff, and their skills. To be sure, the economic crisis itself had both intellectual and institutional aspects and in some sense this distinction can only be analytical. After all, another reason for my interest in think tanks is precisely the convoluted relationship these two dimensions have in them.

The sixth and final premise I want to mention is that think tanks are institutions whose public interventions share many characteristics with those of 'public intellectuals,' and hence can be studied with similar conceptual tools. By this I mean that think tanks and their members could be considered producers of knowledge "in its broadest sense, as communicative ideas that convey cognitive value" (Baert and Shipman 2012: 179). Traditionally, 'public' intellectuals have been understood as those using their influence to address issues concerning the public at large (Collini 2006). Although often put in contrast, I do not distinguish neatly between 'public intellectuals' and 'experts,' inasmuch as public interventions by the latter on the 2008 crisis almost unavoidably have consequences that surpass the field or ambit in which they are considered authorities. However, unlike individual public intellectuals, think tanks require coordination among members who can be working in many policy areas and may have dissimilar, or even conflicting, intellectual dispositions, skills, contacts, and ways of engaging with the public.

From this perspective, one can challenge the allure of 'declinist' theses on the demise of public intellectuals. According to these, traditional public intellectuals in the model of a Bertrand Russell or a Jean Paul Sartre are an endangered species. Beset by a media environment that is generally hostile to their lofty ideals and the breadth of their claims to knowledge, 'declinists' claim intellectuals have become cloistered in an ever more socially detached and politically ineffective academic world (Jacoby 2000; Posner 2003). In their absence, pundits and think-tankers have proliferated (Medvetz 2012b; Misztal 2012). What these theories miss is that, by employing a narrow and overly normative definition of 'intellectual,' restrict prematurely who should be considered as such. Indeed, regardless of how short they fall from an idealised image of the 'public intellectual,' think-tankers often occupy the same space these figures would (see Baert and Booth 2012). Furthermore, as Bauman (1989) noted, determining who qualifies as an 'intellectual' is always an exercise of self-definition. From this standpoint, many of those who subscribe to the declinist view and deride think tanks are, however justified, engaged in boundary work, with a foot in the very category they seek to define.

Ultimately, my purpose is to discern how organisations with intellectual commitments and facing institutional pressures react to a juncture that is unstable and hard to read. As will be borne out throughout the four empirical chapters, even where their public interventions did not betray any questioning of their core tenets—i.e., the 'null' or '*hysteresis*' hypothesis—think tanks underwent noteworthy transformations in how they engage with their publics. The crisis spurred many of them to alter how they presented themselves and through which means, which is a consequence of the challenges they faced when attempting to convince others in a public debate that became ever more fragmented and mistrustful of traditional expert authorities.

More allegorically, this book draws inspiration from Akira Kurosawa's *Rashomon.* In this 1950 film set in mediaeval Japan, four characters—a bandit, a samurai's ghost, his wife, and a woodcutter, the only bystander—describe four different versions of the events surrounding a samurai's murder and the rape of his wife. After these incompatible accounts have been recounted, the spectator is left feeling further from reaching a definite truth, as each narrative 'saves face' for its corresponding narrator. In *Rashomon*, diverging descriptions of putatively the same event evince the pitfalls of relying on only one of them, laying bare the always precarious relationship between the past and how actors approach it. Some scholars have even called this phenomenon the *Rashomon Effect* (see Mazur 1998; Roth and Mehta 2002; Davenport 2010; Anderson 2016).

Nevertheless, there are limits to this approach. Two obvious ones are its *ex post facto* character and its reliance on case-studies rather than a representative sample. More crucially still, this project inescapably depends on the unity of think tanks as organisations. It has been argued that think tanks could be thought of as, rather than distinct 'things,' networks of people and resources closely linked to other institutions. They could be considered 'umbrella organisations' that collect and repackage ideas whose origins lie elsewhere (e.g., parties, universities, interest groups, even internet forums). Without disputing that possibility, this research focuses on particular institutions and their products—rather than on 'travelling ideas' such as austerity—because there are internal reasons why some ideas and not others come to inspire the work of specific organisations. Think tanks have their own logic, otherwise they would not have survived for so long. With their unity in mind, I now delimit how this 'object' is construed throughout the rest of the book.

Think Tanks as a Research Object

Unavoidably, any study on think tanks requires at least a modicum of clarity over what they are in the first place. This has proved to be a surprisingly difficult question that has triggered a long and mostly inconclusive debate (McGann et al. 2014). Medvetz (2012a, b) rightly indicates that this is due to a misplaced emphasis on the issue of independence and a tendency towards 'no true Scotsman' argumentation, producing multiple competing meanings of what a 'true' think tank is. Medvetz sidesteps this difficulty by understanding think tanks as boundary organisations operating across the edges of several fields: they are not a fixed 'thing,' but hybrids of many.

This clarification is theoretically useful, as it suggests think tanks should be studied through what they do rather than what they are. In Medvetz's view, this implies they are defined by their conveyance of tools from diverse fields beyond their traditional setting: the trappings of the Ivy League in Capitol Hill, the buoyancy media demands on how policy papers are drafted, socio-scientific work that is appealing to private benefactors, etc. Some practitioners seem to agree with this emphasis on behaviour rather than on inherent characteristics, as hitherto unavoidable conditions to access the label of think tank—such as non-profit status[2]—are, in certain contexts, beside the point if the institution in question 'behaves' as such.

[2]In parts of Eastern Europe, due to charity regulation and limitations in available funding, many think tanks are registered as for-profit organisations (see Onthinktanks.org 2013).

Nevertheless, defining think tanks as 'boundary organisations' is not specific enough to ascertain what is distinct about their behaviour. It does little to outline the contours of the object or to delimit what should not count as a think tank. After all, there is hardly any organisation that belongs exclusively to one field. To take but one example, even universities are not purely academic endeavours: they have public relations departments, often collaborate with businesses, and their management, academics, and students frequently seek to influence politics and policy. Medvetz (2012a: 128–129) is aware of this objection and submits that think tanks could be considered part of an emergent 'interstitial field.' This take has much to commend to it, but for the purposes of this book, and in more operational terms, I define think tanks by their production of 'public interventions.' More specifically still, think-tankers produce these interventions 'in the name of' an organisation with a history and a name that can be conveyed by many individuals. Hence, as outlined in the following chapters, think tanks can be treated as 'intellectual teams' intervening in the policy debate and operating with the possibilities and constraints derived from being organisations, requiring economic support and coordination.

Yet, should one include international institutions and NGOs such as the OECD, the Bank of International Settlements (Westermeier 2018), or even the IMF, given their think tank-like behaviour, be considered as such? I contend that yes, insofar as they produce non-executive public interventions of relevance to public policy and to the extent they employ 'intervenors' to that effect. They could be said to be partaking in the emerging interstitial field Medvetz speaks about. However, organisations like the above differ from most think tanks in that their work goes beyond producing policy-oriented public interventions. To be sure, the range of activities undertaken by Amnesty International and Human Rights Watch—both curiously listed as think tanks in the University of Pennsylvania *Global Go-To* think tank ranking (McGann 2009 through 2018)—surpasses the production of policy reports.

Could, conversely, a university research centre or department qualify? If researchers of said body cannot speak 'on its behalf,' it should not be treated as a think tank in the sense advanced here, notwithstanding the fact that some have convincingly argued that universities are being pushed in a similar direction (Holmwood 2014). Academics only rarely, and certainly never automatically, intervene in representation of their host institution. The thrust of my understanding of think tanks lies in the fact that they are organisations 'in the name of which' members intervene, similar to the rhetorical device of *prosopopoeia*: in Greek, to *make* a person (or face), and by extension to assign the capacity to speak to something that is itself mute (see González Hernando and Baert forthcoming; see also Cooren 2016).

Given that public interventions are at the core of this definition, I concentrate on them to detect organisational transformations, with an eye on their subject matter, interpretation of events, and format. This exercise is facilitated by the fact that policy reports—that most characteristic of think tank products—often reveal much of the environment in which they are written, as references to scholars, funders, advisors, and policymakers abound. Thus understood, think tanks' public interventions have both material and ideological implications, and 'position' them among like-minded and competing actors.

Furthermore, as many theorists of democracy have insisted (Dewey 1946 [1927]; Rosanvallon 2008), in democratic societies—and arguably in undemocratic ones as well (McGann 2010)—public policy requires, at least perfunctorily, a semblance of being influenced by ideas in order to attain legitimacy. This can be by virtue of either being based on the best available evidence or a robust rationale over what is good or desirable. Given the part played by public interventions as building blocks of the public debate, one can discern the critical role think tanks and similar organisations can have.

The timeframe for this study mostly extends between January 2007 and December 2013, though I also mention some of their work outside this period to provide necessary background or illustrate the direction these think tanks took afterwards. Beginning in 2007 allows noticing if these four think tanks were attentive to the initial signs of instability in the financial system—such as the first write-downs in the US subprime mortgage market and, in Britain, the Northern Rock bank run in September that year. Besides practical considerations, the main reason to stop in 2013 is that by then many of the organisational transformations that derived directly from the crisis were already settled. Some notable junctures within this period are the 2010 parliamentary elections, the austerity programme that ensued, and the growing salience of immigration in the public debate.

Nonetheless, any conceivable timespan would exclude important developments, both from within and outside organisations. In this instance, two of these loom large. The first and most obvious is the Brexit referendum of 2016 and the ensuing debate on the purported emergence of 'post-truth' politics, to which I return in the concluding chapter. The second is the legislative agenda in relation to the campaigning function of charities, leading to changes in Charity Commission (2013) guidelines and the approval of the 'Transparency of Lobbying, Non-Party Campaigning and Trade Union Administration Act' of 2014. Although many of these reforms were dropped by 2016, they had, in their enactment or even their mere possibility, significant effects on the work of think tanks and other third sector organisations.

On this last point, it should be noted that there is no specific legal framework for think tanks in the UK. However, given their 'educational' purposes, the majority are registered as charities,[3] benefitting from tax exceptions that would otherwise render many of them unviable. As such, most fall under the supervision of the Charity Commission. This means they have a board of trustees and must justify their provision of 'public benefit,' which establishes limits to the campaigning activities they can undertake and precludes any direct party-political role, especially in the period before elections. However, some think tanks, or parts of them, are not charities but considered limited-liability companies registered in the Companies House (see Chapter 4).

These legal requirements might explain why some think tanks, such as IEA and Civitas, declare not to have a 'corporate view.' However, for the purposes of this book, the very fact they publish certain experts and not others can be telling of what they consider relevant additions to the policy debate. For instance, the IEA, given its history, would be unlikely to release a report arguing for the nationalisation of industries;

[3]This is also the case in the US, where most file as tax-exempt 501(c)(3) non-profits.

its consent to publishing external authors with their logo is implicitly a sanctioning of their proposals. The 'brand' of a think tank, in this sense, has a weight, and is associated in networks of ideas and people. To introduce these 'brands' for the cases under scrutiny, I close this chapter by presenting the four think tanks I focus on and why.

Which Think Tanks?

The latest edition of the most prestigious global think tank ranking (McGann 2018) claims there are 444 such organisations in Britain, up from 288 the year before (McGann 2017), which places the UK only behind the US and China. Although in any such exercise, definitions and methodology are bound to be contentious, this gives a good illustration of the copious number of think tanks and akin institutions in Britain. Since describing in any depth the institutional and intellectual responses to the crisis of that many cases is untenable and, given that many concentrate in narrow policy areas, I opt instead to focus on as small a number and as diverse a sample as possible. To that end, although I do not want to provide a taxonomy of think tanks, a minimum explanation of the selection criteria is required.

I have chosen case-studies in consideration of their potential theoretical yield for the comparative analysis that follows, guided by three variables. The first relates to their professed ideological views, their claims to defend particular ideas of society and the economy, especially in relation to economic deregulation. Many have written on the biases and interests that fuelled the advent of these organisations in the 1970s, and the degree to which they have been intertwined with corporate, ideological, and political interests (Cockett 1995; Muller 1996; Stone 1991; Jacques et al. 2008). After all, one can associate most think tanks with political positions that exist elsewhere (e.g., conservative, libertarian, social-democrat). Furthermore, and it almost goes without saying, views on how to organise an economy are linked to different positions in the political spectrum, and thus are likely to be associated with different social relationships. That is, most think tanks can be understood to be, roughly speaking, part of the left or the right and their networks. Of course, this is a contextual variable. In Britain, in mainstream terms at least, the political left is connected with the promotion of the welfare state and the right with *laissez-faire* market liberalism. Hence, think tanks associated with the left are more likely than those on the right to receive substantive support from trade unions, the opposite being the case for corporate donations. Albeit these categories are vast and internally diverse, they grant a starting point from with to distinguish further.

The second sampling variable is the perceived proximity of a think tank to networks of political power. Some of the most influential think tanks are informally linked to parties (or one of its factions), even if not legally allowed, being charities, to be overtly party-political. Expectably, being perceived to share an intellectual position with a party that might form a government is not enough to claim a think tank exerts any policy influence, but it does affect its strategy and target audience. This is critical for what type of public interventions a think tank is likely to produce, especially in their direct targeting of policymakers or larger publics—what in the Discursive Institutionalism literature is called their reliance on, respectively, either 'coordinative' or 'communicative' discourse (Schmidt 2008). Conversely, even if think tanks wish to be seen as influential, it can be an advantage to maintain a perceived distance from politics. This helps an organisation to seem independent enough to be perceived as having cognitive autonomy, allowing it to give policymakers the expert endorsement that more partisan institutions would be unable to grant. On the other hand, a think tank not directly linked to influential politicians can often be more outspoken than those who wish to seem 'middle of the road,' which in turn might raise their visibility, if perhaps at the expense of their reputation for impartiality.

Relatedly, the third variable distinguishes between a propensity to argue from a normative standpoint or as neutral specialists. Even conceding that the traditional institutional sources of expert authority have eroded (Brooks 2012) there is still a relevant distinction to be made between those who attempt to seem politically and ideologically impartial or otherwise (Baert and Booth 2012). Of course, while even overtly partisan institutions claim some form of expertise—which often involves a delicate balancing act—other think tanks position themselves as experts precisely by virtue of being seen as neutral, cross-party, or scholarly. In that sense, one could hypothesise that think-tankers who wish their organisation to be seen as non-partisan and guided by socio-scientific evidence will value their financial and political autonomy and keep a certain distance from the political fray. The price of this position is, in general, a measure of intellectual separation that hampers open participation in the political tug of war, and the need for a minimum detachment from the media, political parties, interest groups, and many non-academic sources of research funding. Thus, seeming 'neutral' is a constant endeavour, necessitating a degree of open-endedness and academic resources that not everyone can afford.

With the above in mind, I decided to concentrate on four British think tanks. They were chosen to comprise the widest array of engagement strategies (e.g., neutrality, advocacy), ideological positions (pro and against economic deregulation), and types of relationship to political parties (direct or distant) in such few examples, sacrificing as little scope as possible for the sake of depth. Crucially, all are 'generalists,' as opposed to specialists producing work on a narrow policy area, mainly because an important dimension of a think tank's response to the crisis is precisely their choice of what to inquire into.

The first is the left-of-centre New Economics Foundation (NEF). It was founded in 1986 to offer innovative economic advice with an emphasis on sustainability and wellbeing. Some of its favoured policy areas include environmental policy, local economies, and welfare, but produces research in various others, including work, happiness, and democracy. It is a media-friendly think tank that greatly depends on charitable funding. Since NEF's core ideas include de-growth (i.e., GDP growth cannot continue ad infinitum in a finite planet), heterodox economics, and their opposition to the dominance of finance, presumably the aftermath of 2008 was a propitious moment for the dissemination of its message.

The second is one of the historical bastions of the British free-market movement, the Adam Smith Institute (ASI). Founded in 1977, the ASI is characterised by its outspoken opposition to regulation, taxation, state provision, and public ownership. The ASI has a long tradition of media savviness, carried forward to this day by a small and young team intervening tirelessly through reports, op-eds, blogs, and social media.

A second wave think tank mostly supported by individual and corporate donors, the ASI famously helped lay the intellectual groundwork for the Thatcher reforms of the 1980s. As it promotes market liberalisation and relies on anonymous private donations, the 2008 crisis presented both important threats and opportunities to this most emblematic of think tanks.

The third is the National Institute for Economic and Social Research (NIESR). Founded in 1938, NIESR's primary objective is to produce rigorous social science research, especially in applied economics, to provide sound independent advice to government. An important part of NIESR's research outputs are economic forecasts and reports on issues broadly concerning growth and productivity, such as employment, training, inequality, and immigration. NIESR relies mostly on research contracts and subscriptions to its econometric model, mixing commissioned projects and periodic publications such as economic reviews and academic journals. Their attempts to be seen as neutral and objective, parallel to greater attention and scrutiny of their work, makes their experience of the crisis fascinating.

The fourth and last is Policy Exchange (PX). Founded in 2002 by Conservative MPs, it is sometimes considered the most influential single institution of the Cameron era (Pautz 2012b). PX's work is based on localism, voluntarism, and free-market solutions to a wide range of policy problems, including housing, healthcare, education, security, and economic policy. PX is supported by many sources of private and charitable funding and is strongly associated with Tory modernisers. As such, PX was seminal in informing the ideas and policies of quite a few influential politicians, both while in opposition and while in government. Hence, not only are PX's policy proposals consequential on account of who is likely to listen to them, but also because they are illustrative of the parallel trajectories of the Conservative party under Cameron and of a think tank tightly linked to it.

The remainder of this book is structured as follows. The next chapter delves deeper into theoretical approaches to think tanks, intellectuals, and policy change. It expands on my intended contribution to this literature, as well as on the methodology I employ going forward. Readers who are less interested in these scholarly debates are invited to continue directly with the four chapters that ensue—though please do take notice of the *hysteresis* hypothesis model (Fig. 2.1). Each of the following empirical chapters describes the institutional characteristics of one of these think tanks, traces the changes they underwent after the crisis, and draws general conclusions on how these organisations operate. The last chapter compares the fates of these think tanks and reflects on what the transformations in their mode of public engagement reveal of broader developments in British policymaking.

References

Abelson, D. (2002). Do think tanks matter? Opportunities, constraints and incentives for think tanks in Canada and the United States. *Global Society, 14*(2), 213–236.

Abelson, D. (2012). Theoretical models and approaches to understanding the role of lobbies and think tanks in US foreign policy. In S. Brooks, D. Stasiak, & T. Zyro (Eds.), *Policy expertise in contemporary democracies*. Farnham: Ashgate.

Abolafia, M. (2010). Narrative construction as sensemaking: How a central bank thinks. *Organization Studies, 31*(3), 349–367.

Anderson, R. (2016). The Rashomon effect and communication. *Canadian Journal of Communication, 41*(2), 250–265.

Aupers, S. (2012). Trust no one: Modernization, paranoia and conspiracy culture. *European Journal of Communication, 27*(22), 22–34.

Baert, P., & Booth, J. (2012). Tensions within the public intellectual: Political interventions from Dreyfus to the new social media. *International Journal of Politics, Culture and Society, 25*(4), 111–126.

Baert, P., & Shipman, A. (2012). Transforming the intellectual. In P. Baert & F. Domínguez Rubio (Eds.), *The politics of knowledge* (pp. 179–204). New York: Routledge.

Ball, S., & Exley, S. (2010). Making policy with 'good ideas': Policy networks and the 'intellectuals' of New Labour. *Journal of Education Policy, 25*(2), 151–169.

Banet-Weiser, S. (2012). Branding the crisis. In J. Caraça, G. Cardoso, & M. Castells (Eds.), *Aftermath: The cultures of economic crisis* (pp. 107–131). Oxford: Oxford University Press.

Bauman, Z. (1989). *Legislators and interpreters: On modernity, postmodernity and intellectuals*. Cambridge: Polity.

BBC. (2017). Crash was economists' 'Michael Fish' moment, says Andy Haldane. Accessed 15 September 2018. https://www.bbc.com/news/uk-politics-38525924.

Beck, U., & Wehling, P. (2012). The politics of non-knowing: An emergent area of social and political conflict in reflexive modernity. In P. Baert & F. Domínguez Rubio (Eds.), *The politics of knowledge* (pp. 33–57). New York: Routledge.

Bentham, J. (2006). The IPPR and Demos: Think tanks of the new social democracy. *Political Quarterly, 77*(2), 166–174.

Berry, M. (2016). No alternative to austerity: How BBC broadcast news reported the deficit debate. *Media, Culture and Society, 38*(6), 844–863.

Boin, A., t'Hart, P., & McConnell, A. (2009). Crisis exploitation: Political and policy impacts of framing contests. *Journal of European Public Policy, 16*(1), 81–106.

Boltanski, L., & Thévenot, L. (2006). *On justification: Economies of worth*. Princeton: Princeton University Press.

Bourdieu, P. (1988). Vive la crise!: For heterodoxy in social science. *Theory & Society, 17*(5), 773–787.

Brändström, A., & Kuipers, S. (2003). From 'normal incidents' to political crises: Understanding the selective politicization of policy failures. *Government and Opposition, 38*(3), 279–305.

Brooks, S. (2012). Speaking truth to power: The paradox of the intellectual in the visual information age. In S. Brooks, D. Stasiak, & T. Zyro (Eds.), *Policy expertise in contemporary democracies* (pp. 69–85). Farnham: Ashgate.

Bryan, D., Martin, R., Montgomerie, J., & Williams, K. (2012). An important failure: Knowledge limits and the financial crisis. *Economy & Society, 41*(3), 299–315.

Campbell, J. (2002). Ideas, politics and public policy. *Annual Review of Sociology, 28*, 21–38.

Charity Commission. (2013, September 1). *What makes a charity (CC4)*. Accessed 10 January 2016. https://www.gov.uk/government/publications/what-makes-a-charity-cc4.

Cockett, R. (1995). *Thinking the unthinkable: Think tanks and the economic counter-revolution 1931–1983*. London: HarperCollins.

Collini, S. (2006). *Absent minds: Intellectuals in Britain*. Cambridge: Cambridge University Press.

Cooren, F. (2016). *Organizational discourse: Communication and constitution*. London: Wiley.
Crehan, K. (2011). Gramsci's concept of common sense: A useful concept for anthropologists? *Journal of Modern Italian Studies, 16*(2), 273–287.
Crehan, K. (2016). *Gramsci's common sense: Inequality and its narratives*. Durham: Duke University Press.
Crouch, C. (2011). *The strange non-death of neoliberalism*. Cambridge: Polity.
Davenport, C. (2010). *Media bias, perspective, and state repression: The Black Panther Party*. Cambridge: Cambridge University Press.
Davies, W., & McGoey, L. (2012). Rationalities of ignorance: On financial crisis and the ambivalence of neoliberal epistemology. *Economy & Society, 41*(1), 64–83.
de Goede, M. (2009). Finance and the excess: The politics of visibility in the credit crisis. *Zeitschrift für Internationale Beziehungen, 16*(2), 295–306.
Denham, A., & Garnett, M. (1998). *British think tanks and the climate of opinion*. London: UCL Press.
Denham, A., & Stone, D. (Eds.). (2004). *Think tank traditions*. Manchester: Manchester University Press.
Desai, R. (1994). Second hand dealers in ideas: Think tanks and Thatcherite hegemony. *New Left Review, 203*(1), 27–64.
Dewey, J. (1946 [1927]). *The public and its problems: An essay in political inquiry*. Chicago: Gateway Books.
Domhoff, G. W. (1967). *Who rules America?* New York: McGraw Hill.
Engelen, E., Erturk, I., Froud, J., Johal, S., Leaver, A., Moran, M., et al. (2011). *Misrule of experts? The financial crisis as elite debacle* (CRESC Working Paper Series, 94).
Eyal, G., & Pok, G. (2011). From a sociology of professions to a sociology of expertise. Accessed 20 February 2015. http://cast.ku.dk/papers_security_expertise/Eyal__2011_From_a_sociology_of_professions_to_a_sociology_of_expertise.pdf.
Eyerman, R. (2011). Intellectuals and cultural trauma. *European Journal of Social Theory, 14*(4), 453–467.
Fischer, K., & Plehwe, D. (2013). Redes de think tanks e intelectuales de derecha en América Latina. *Nueva Sociedad, 245*, 70–86.
Fourcade, M. (2009). *Economists and societies: Discipline and profession in the United States, Britain and France, 1890s to 1990s*. Princeton: Princeton University Press.
Gamble, A. (2009). *The spectre at the feast: Capitalist crisis and the politics of recession*. Basingstoke: Palgrave Macmillan.

Gills, B. (2010). The return of crisis in the era of globalization: One crisis, or many? *Globalizations, 7*(1–2), 3–8.

Gramsci. A. (1999 [1971]). *Selections from the prison notebooks*. London: Elecbooks.

González Hernando, M., Pautz, H., & Stone, D. (2018). Think tanks in 'hard times': The global financial crisis and economic advice. *Policy & Society, 37*(2), 125–139.

González Hernando, M., & Baert, P. (forthcoming). Collectives of intellectuals: Their cohesiveness, accountability, and who can speak on their behalf. *The Sociological Review*.

Herbst, S. (2003). Political authority in a mediated age. *Theory & Society, 32*(4), 481–503.

Holmwood, J. (2014). Sociology's past and futures: The impact of external structure, policy and financing. In J. Holmwood & J. Scott (Eds.), *A handbook of British sociology*. London: Palgrave.

Jacoby, R. (2000). *The last intellectuals: American culture in the age of academe*. New York: Basic Books.

Jacques, P., Dunlap, R., & Freeman, M. (2008). The organisation of denial: Conservative think tanks and environmental scepticism. *Environmental Politics, 17*(3), 349–385.

James, S. (1993). The idea brokers: The impact of think tanks on British government. *Public Administration, 71*, 491–506.

Kay, L., Smith, K., & Torres, J. (2013). Think tanks as research mediators? Case studies from public health. *Evidence and Policy, 59*(3), 371–390.

Krastev, I. (2001). Think tanks: Making and faking influence. *Southeast European and Black Sea Studies, 1*(2), 17–38.

Lawson, T. (2009). The current economic crisis: Its nature and the course of academic economics. *Cambridge Journal of Economics, 33*(4), 759–777.

Lischinsky, A. (2011). In times of crisis: A corpus approach to the construction of the global financial crisis in annual reports. *Critical Discourse Studies, 8*(3), 153–168.

Lo, A. (2011). *Reading about the financial crisis: A 21-book review*. Social Science Research Network. Accessed 15 March 2013. http://ssrn.com/abstract=1949908.

Lucas, R. (2003). Macroeconomic Priorities. *American Economic Review, 93*(1), 1–14.

Mazur, A. (1998). *A hazardous inquiry: The Rashomon effect at Love Canal*. Cambridge, MA: Harvard University Press.

McGann, J. (2009). *2008 global go to think tanks and policy advice ranking.* Think Tanks and Civil Society Program. University of Pennsylvania.
McGann, J. (2010). *Democratization and market reform in developing and transitional countries: Think tanks as catalysts.* London: Routledge.
McGann, J. (2017). *2016 global go to think tanks and policy advice ranking.* Think Tanks and Civil Society Program. University of Pennsylvania.
McGann, J. (2018). *2017 global go to think tanks and policy advice ranking..* Think Tanks and Civil Society Program. University of Pennsylvania.
McGann, J., Viden, A., & Rafferty, J. (Eds.). (2014). *How think tanks shape social development policies.* Philadelphia: University of Pennsylvania Press.
McGoey, L. (2012). Strategic unknowns: Towards a sociology of ignorance. *Economy & Society, 41*(1), 1–16.
McLennan, G. (2004). *Dynamics of transformative ideas in contemporary public discourse, 2002–2003.* Accessed 15 October 2013. http://www.esds.ac.uk/doc/5312/mrdoc/pdf/q5312uguide.pdf.
McNutt, K., & Marchildon, G. (2009). Think tanks and the web: Measuring visibility and influence. *Canadian Public Policy, 35*(2), 219–236.
Medvetz, T. (2012a). *Think tanks in America.* Chicago: University of Chicago Press.
Medvetz, T. (2012b). Murky power: 'Think tanks' as boundary organizations. In D. Golsorkhi, D. Courpasson, & J. Sallaz (Eds.), *Rethinking power in organizations, institutions, and markets: Research in the sociology of organizations* (pp. 113–133). Bingley: Emerald Group Publishing.
Mills, C. W. (1956). *The power elite.* Oxford: Oxford University Press.
Misztal, B. (2012). Public intellectuals and think tanks: A free market in ideas? *International Journal of Politics, Culture and Society, 25*(4), 127–141.
Morin, E. (1976). Pour une crisologie. *Communications, 25,* 149–163.
Muller, C. (1996). The institute of economic affairs: Undermining the post-war consensus. *Contemporary British History, 10*(1), 88–110.
Onthinktanks.org. (2013). *For-profit think tanks and implications for funders.* Accessed 25 March 2015. https://onthinktanks.org/articles/for-profit-think-tanks-and-implications-for-funders/.
Pautz, H. (2012a). *Think tanks, social democracy and social policy.* London: Palgrave Macmillan.
Pautz, H. (2012b). The think tanks behind 'cameronism'. *British Journal of Politics and International Relations, 15*(3), 362–377.
Pautz, H. (2016). Managing the crisis? Think tanks and the British response to global financial crisis and great recession. *Critical Policy Studies, 11*(2), 191–210 [Online early access].

Pierson, R. (1994). The epistemic authority of expertise. *PSA: Proceeding of the Biennial Meeting of the Philosophy of Science Association, 1*, 398–405.
Plehwe, D. (2010). Think tanks und Entwicklung. Bessere Integration von Wissenschaft und Gesellschaft? *Journal für Entwicklungspolitik, 26*(2), 9–37.
Posner, R. (2003). *Public intellectuals: A study of decline*. Cambridge, MA: Harvard University Press.
Rantanen, T. (2012). In nationalism we trust? In J. Caraça, G. Cardoso, & M. Castells (Eds.), *Aftermath: The cultures of economic crisis* (pp. 132–153). Oxford: Oxford University Press.
Rohloff, A., & Wright, S. (2010). Moral panic and social theory: Beyond the heuristic. *Current Sociology, 58*(3), 403–419.
Roitman, J. (2013). *Anti-crisis*. London: Duke University Press.
Rosanvallon, P. (2008). *La légitimité démocratique: Impartialité, réflexivité, proximité*. Paris: Seuil.
Roth, W., & Mehta, J. (2002). The Rashomon effect: Combining positivist and interpretivist approaches in the analysis of contested events. *Sociological Methods and Research, 31*(2), 131–173.
Sandvoss, C. (2010). Conceptualizing the global economic crisis in popular communication research. *Popular Communication: The International Journal of Media and Culture, 8*(3), 154–161.
Schlesinger, P. (2009). Creativity and the experts: New Labour, think tanks, and the policy process. *International Journal of Press/Politics, 14*(1), 3–20.
Schmidt, V. (2008). Discursive institutionalism: The explanatory power of ideas and discourse. *Political Science, 11*(1), 303–322.
Schmidt, V., & Thatcher, M. (2013). *Resilient liberalism in Europe's political economy*. Cambridge: Cambridge University Press.
Scott Solomon, M. (2010). Critical ideas in times of crisis: Reconsidering Smith, Marx, Keynes, and Hayek. *Globalizations, 7*(1–2), 127–135.
Sinclair, T. (2010). Round up the usual suspects: Blame and the subprime crisis. *New Political Economy, 15*(1), 91–107.
Stone, D. (1991). Old guard versus new partisans: Think tanks in transition. *Australian Journal of Political Science, 26*(2), 197–215.
Stone, D. (1996). From the margins of politics: The influence of think-tanks in Britain. *West European Politics, 19*(4), 675–692.
Stone, D. (2007). Recycling bins, garbage cans or think tanks? Three myths regarding policy analysis institutes. *Public Administration, 85*(2), 259–278.
t'Hart, P., & Tindall, K. (Eds.). (2009). *Framing the global economic downturn: Crisis rhetoric and the politics of recessions*. Sydney: ANU Press.

Tesseyman, A. J. (1999). *The new right think tanks and policy change in the UK.* (PhD thesis). Department of Politics, University of York.
Thompson, J. (2012). The metamorphosis of a crisis. In J. Caraça, G. Cardoso, & M. Castells (Eds.), *Aftermath: The cultures of economic crisis* (pp. 59–81). Oxford: Oxford University Press.
Tooze, A. (2019). *Crashed: How a decade of financial crises changed the world.* London: Penguin.
Walby, S. (2016). *Crisis.* Cambridge: Polity.
Westermeier, C. (2018). The Bank of International Settlements as a think tank for financial policy-making. *Policy & Society, 37*(2), 170–187.
Wren-Lewis, S. (2018). *The lies we were told: Politics, economics, austerity and Brexit.* Bristol: Bristol University Press.

2

How Thinking Takes Place in Think Tanks

This chapter lays the theoretical and methodological groundwork for this book. I start by presenting and examining three approaches that inform the conceptual apparatus I use going forward: Medvetz's employment of Bourdieusian field theory; the sociology of intellectual interventions; and Neoinstitutionalist studies of organisations and public policy. I then expand on how I use think tanks' public interventions as indexes of intellectual and institutional change, as well as on my approach to interviews. These reflections give form to the research design that structures the following chapters, centred on the 'null,' *hysteresis* hypothesis.

Think Tanks as Boundary-Crossers

Thomas Medvetz's (2012) *Think Tanks in America* is perhaps the most influential recent book on the topic. It challenges a tendency in the scholarship on think tanks to fixate on definitions and typologies. Concerning the former, Medvetz claims there has been a proclivity for tautological argumentation in relation to what an organisation must be

M. González Hernando, *British Think Tanks After the 2008 Global Financial Crisis*, Palgrave Studies in Science, Knowledge and Policy, https://doi.org/10.1007/978-3-030-20370-2_2

independent from to be considered a 'true' think tank—universities, government, special interests, etc. He argues that believing think tanks should necessarily be detached from the state, for example, rules out most comparable organisations in Germany and China, without much theoretical insight gained from that exclusion. Regarding typologies (Weaver 1989), Medvetz critiques their proneness to pigeonhole think tanks too neatly and exaggerate their stability.

To tackle these shortcomings, Medvetz proposes a model based on Pierre Bourdieu's sociology, especially by employing the concepts of 'field' and 'capital'—or in simpler terms, areas of contention with their own rules and hierarchies (e.g., academia) and resources associated with them (e.g., a doctorate and the prestige it bestows). Medvetz understands think tanks as boundary institutions partaking in at least four fields and employing their respective capitals—academia, politics, the media, and the economic world. Think tanks thus perform a balancing act between a broad set of aims and resources (see Bourdieu and Wacquant 1992; Couldry 2003). By this definition, they carry with them extraneous forces into each field they enter in contact with. Think tanks bring economic considerations into socio-scientific research, promote academic ideas through the media, commission scholars and businesspeople to write policy reports, etcetera.

However, Medvetz ponders, how can such hybrid organisations survive and thrive without being defeated in every field by more specialised competition? His answer is that think tanks' competitive advantage lies in the fact that their capacity to cross boundaries grants them a measure of plasticity and resilience. For instance, a think tank with considerable media visibility could, if needed, give politicians a platform, earn their favour, and in the process become policy relevant; another think tank with robust links with business might pursue academic gravitas to bolster the credibility of their proposals.

These ideas have two crucial corollaries. First, the efforts of a think tank to accumulate a specific type of capital might weaken its ability to acquire others. An organisation that invests heavily in its academic profile could be perceived as too esoteric for the business world, too stuffy for the media, and politically ineffectual; another too strongly associated with its economic benefactors could be accused of obscurity and bias by observers from the academic and media world, and seen as a liability in the political field. That is why for Medvetz the best course of action is most frequently to find a middle ground, acquiring greater volumes of each of the four capitals while avoiding unbalance; remaining halfway between fields that can be suspicious of each other. Second, the same factors that undermine think tanks' cognitive autonomy can operate as curious sources of flexibility and power. By virtue of their hybrid nature, think-tankers are more politically savvy than most academics, more scholarly than most politicians, more media-friendly than most lobby groups, etc.

For Medvetz, the rise of think tanks in the United States began in earnest around the time of the Vietnam war, with the emergence of critiques of technocracy, positivism, and academia coming from both the left and the right. Accusing universities and an earlier generation of academically oriented policy institutes of being elitist, untimely for policy, self-interested, and obscure, the US equivalents of Denham and Garnett's second wave sought to carve a space for themselves in an increasingly crowded and adversarial policy debate. In the process, Medvetz claims these organisations blurred the boundaries between expertise and politics, hampered the possibilities of more autonomous intellectuals to thrive independently, and produced vast volumes of degraded forms of socio-scientific knowledge—stockpiles of papers whose conclusions were settled long before any research was conducted.

One of this theory's crucial advantages lies in its avoidance of an over-deterministic view of think tanks without portraying them as completely autonomous either—they respond to many, sometimes conflicting pressures. Another of its accomplishment is that it broadens the possible aims of think tanks beyond the pursuit of academic rigour. A think tank's audiences may have different readings from the same policy report, and assuming that any of them is privileged (e.g., that of academia) might not reflect the actual objectives of these institutions. In that line, Medvetz contends that different think tanks value cognitive autonomy differently. For instance, changing one's policy position due to socio-scientific evidence might be construed in some quarters as betraying one's ideals, which could jeopardise the support they receive from ideological allies.

My own thinking is heavily inspired by Medvetz, and one could say this book is my attempt to expand on his framework. I seek that by examining a different context to answer a different question: instead of tracing how think tanks emerged at the margins of fields, I am interested in how they reacted to a crisis that was, at the same time, economic, political, and epistemic—not to mention the momentous changes the media field was also undergoing. In the process, I also hope to refine some theoretical aspects that are not directly addressed in Medvetz's book, particularly in relation to the dynamics of the fields in which think tanks participate.

With that in mind, I would like to highlight an issue that, although implicit in *Think Tanks in America*, is explored only cursorily. Here some recourse to Bourdieusian terminology is necessary. Medvetz claims think tanks operate in at least four fields, garnering and deploying their corresponding forms of capital: the economy (money and the means to acquire it), politics (power and attention from the powerful), the media (visibility), and academia (scholarly credibility).[1] Following Bourdieu, each of these fields is governed by a specific logic while engaged in a struggle over their remit and power in relation to other fields. That is, in every field and for every capital there are at least two types of dispute: one 'external' and one 'internal.'

By external dispute I mean the contested worth of different types of capital in determining the pecking order within any given field. Most Bourdieusian theorists cover these matters through the distinction between more 'autonomous' and more 'heteronomous' fields, the degree to their which internal competition is affected by external factors (e.g., Pels 1995; Benson 1999). One way of measuring this is through 'capital convertibility,' the extent to which resources that are largely extraneous to a field play a part in the results of its 'game' (Calhoun 1995). For instance, who is best positioned in a relatively autonomous field (e.g., mathematics) will generally, or at least ideally, have little to do with who has the most money or political contacts, especially in comparison to a more heteronomous one

[1] Savage et al. (2005) claim Bourdieu's concept of capital expanded from four types—economic, cultural, social, symbolic—towards more field-specific forms—e.g., educational—in his later work.

(e.g., public policy). It follows that part of what think tanks do, situated as they are in in-between spaces, is to facilitate capital conversion: they wield academic credentials and ideas to justify a certain political course of action, invest economic resources to fund particular research outcomes,[2] etc. Put this way, heteronomy is built into their very core.

Yet, the process of 'converting' one form of capital into another is never unproblematic. Those who have invested heavily in a particular form of capital profit from making it recognised ever more widely. For instance, if one has vast sums of money, one stands to benefit if more and more things become purchasable, and if one has strong academic credentials, one has an interest in promoting a greater role for experts and the university in society. But those efforts to convert capitals are often met with scepticism by those with a different 'portfolio.' A corollary of this suspicion is that there is friction in any capital conversion, expressed in the dynamics of 'loss' and 'concealment' (Bourdieu 1986: 54). For instance, those who possess strong scholarly credentials will likely argue for a greater influence of science and academia in policymaking, frequently involving a volume loss: it is unlikely that someone with high levels of academic capital could seamlessly turn it, to the same degree, into political power. Conversely, those with vast financial resources seeking policy influence cannot, solely by virtue of this possession, openly command political clout, and hence economic capital in politics is usually 'concealed' in other forms. This is why Medvetz claims think tanks are boundary institutions that transform the contours of the fields in which they operate—one could add, by facilitating capital conversion.

So far we have not drifted a long way from Medvetz's argument—indeed, in a later publication he refers explicitly to 'capital conversion rates' (Medvetz 2015: 232). But these issues also lead to the second type of field dispute I mentioned earlier. That is, how actual resources, whatever their kind, are understood as capitals that can be mobilised within a field in the first place (Savage et al. 2005). Bourdieu defines capital as 'accumulated labour' that can be embedded in agents (e.g., their

[2] The IEA's 2014 Brexit Prize—an essay competition seeking the best proposal for a policy mechanism to facilitate Britain's potential exit from the EU—is a good example of such an attempt. See (accessed 20 May 2015) https://iea.org.uk/brexit.

education, their demeanour) or transmitted through objects, institutions, symbols, or social relations (Bourdieu 1986: 46). Implicit in that definition is a requirement for these capitals to be valued by others, or in other words, that said accumulated labour is acknowledged as such in relation to that of others.

The most common way to understand capital is, of course, money. In that case, the means by which relative wealth or want is measured are relatively straightforward, being this form of capital, at least in contemporary capitalist societies, abstract, quantifiable, ubiquitous yet scarce, and highly convertible. Indeed, it is perhaps said convertibility what grants the economic field its specific power (Bourdieu 2000). Furthermore, even when a form of economic capital—e.g., a currency—changes in value, its fluctuations are generally expressible in a calculable 'exchange rate.' When the price of a stock or currency oscillates, notwithstanding uncertainty over its future worth, there is at any point in time a numerical measure by which it can be compared to (and exchanged with) other economic resources.

To be sure, most forms of capital are not as easy to quantify as money, and their convertibility is not as seamless. Even economic capital is less transparent than commonly assumed, as studies on the performativity of market devices have shown (Muniesa 2014). Be that as it may, one common way of measuring academic, media, or political capital is, similarly, to count whatever can be counted. For instance, to assess the academic capital of a think tank one could report the proportion of its employees who possess a doctorate, how much its publications are cited by peers and experts, etc. However, this is limited and by no means the only way. Since a capital is firstly a resource that has to be acknowledged as such in social interaction and whose worth varies over time, it is necessary to explore these issues from a qualitative and historical perspective. Otherwise, one risks neglecting tensions within the academic field that have important effects on think tanks. For example, it is not enough to say that a policy institute employs many economists. Within that discipline, different schools (New Keynesian, Austrian, Monetarist, etc.) structure their own networks (departments, journals, associations, etc.) that compete for prestige and attention and have differing views of what 'good economics' is.

Crucially for this book, these internal tensions are even more pressing in moments when the worth of resources and the boundaries of fields are in flux. Following the example of economics, after the financial crisis, the ideas of scholars who were hitherto relatively marginal (e.g., Hyman Minsky) gained traction almost overnight, while those of the dominant (e.g., Robert Lucas) faced important challenges. Furthermore, even when we speak of economic capital—perhaps the most seemingly transparent of all kinds—its provenance and position is never purely economic. To receive funding from a research council grant is quite different from doing so as a donation from a wealthy individual, even if at the same volume. Not only are the 'strings' it comes attached with different, but so are its reputational effects.

These considerations are partly inspired by Michel Dobry's (2009) *Sociologie des crises politiques*, a book which unfortunately has had only limited circulation outside French-speaking academia. Dobry's sociology—also influenced by Bourdieu's field theory, and initially applied to Eastern European countries in the immediate aftermath of the fall of the Iron Curtain—contends that during political crises the worth of resources becomes unstable. That is because in such moments the boundaries of social fields—or, in his words, sectors—become porous and actors are made aware of how their actions reverberate beyond their usual sphere of influence. These broader effects, although in some sense always present, often go unnoticed in periods of 'normality,' where social action is more readily circumscribed in their purported remit. Dobry speaks of crises as contexts of *interdépendance tactique élargie*, in which the 'tactical interdependence' of social action is 'widened' across spheres of society that are traditionally thought of as separate. For instance, during a financial crisis, economic theories, as technical and objective as their authors might wish them to be, cannot avoid being seen as political.

However, warns Dobry, one should not overstate the distinction between crises and normality. Citing Nietzsche,[3] he claims the former reveals hidden aspects of the latter, rather than transfigure it. Economics

[3] "The value of all morbid states consists in that they show under a magnifying glass certain conditions that, although normal, are not easily visible in their normal state" (Nietzsche, in Dobry 2009: 317 [my translation]).

is always political, after all. Nor should one overstate the capacity of individuals to influence the outcome of a crisis once the 'field is open'; a mistake he calls the *illusion héroïque*. Although new agents might rise in prominence, the efficacity of their actions depends on the conduits through which they are conveyed to a broader network, access to which often continues to be brokered by established elites. Indeed, much of the old Communist guard remained politically central in Eastern Europe well after 1989. For the above reasons, I argue that Medvetz's theory should be expanded to consider the instability of fields and capitals. Recognising that capitals are variegated and everchanging and that fields are unstable points to a juncture where Medvetz's model needs to be 'unsettled.'

In this line, Medvetz (2015: 222) claimed think tanks can exert, in their capacity to broker at the intersection of fields, an influence akin to Bourdieu's 'symbolic power'—"the power to impose meaning and legitimate distinctions"—delimiting what constitutes a valuable capital. And to be sure, brokering is never done from a neutral position. For instance, political capital is not a strictly one-dimensional affair, in one end of the scale there being 'more' and in another 'less.' The political field is not as simple, and something similar also applies to other fields. The way how scholarly credentials are used by think tanks to assert credibility can be itself strategic, and the promotion of certain economic theories already implies a view of what is sound economics. In that sense, the 'worth' of different academic resources is impossible to determine without acknowledging that any such measure is done from a vantage-point, or what is sometimes in Bourdieusian theory called a *habitus*.[4] Without this in mind, one could mistakenly think of fields as spaces where the rules and stakes of the game are immutable and clear to all.

[4]Interestingly, in Medvetz's book the concept of *habitus* seldom appears—even though it is central in Bourdieusian thought and often taught in sociology courses alongside 'field' and 'capital.' It does so only when discussing the agency of policy experts but never in relation to think tanks themselves (Medvetz 2012: 153–155). As there have been some attempts to employ *habitus* to organisations (Emirbayer and Johnson 2008; Swartz 2013), I find that absence intriguing, if justifiable on two counts: because of this concept's focus on early socialisation; and because employing *habitus* at an organisational level risks neglecting that of its members.

These reflections are important because without a concept that grounds the position of a think tank one is left wondering what explains their ideological resilience. Why would they not, rather than defend a set of ideas that might be on the wane, associate themselves with whatever theory has the most purchase in economics? By way of illustration, after 2008 free-market theories came under attack, at least for a while, which could put in a predicament those who have invested heavily on them. Nonetheless, and unsurprisingly, think tanks are unlikely to simply sell their shares and cut their losses. There is remarkable sensibility to change in Medvetz's approach to think tanks, but without explicit references to their position and disposition within fields there is little space to explain the firmness of their intellectual and institutional commitments.

Of course, theories of stability abound. Neoinstitutionalism (covered later in this chapter) is a good example, and Bourdieu himself is often considered a theorist of social reproduction rather than of change (Gorski 2013). Pertinently for this book, in the context of transformations of academia in France, Bourdieu (1990) used the concept of *hysteresis*—the lagging off of an effect beyond the persistence of its cause—to explain the persistence of ways of looking at the world in the face of extreme social change (Kerr and Robinson 2009). In the context of this research, *hysteresis* implies that the ideas produced by a think tank are likely to be similar to those they have advocated for in the past. Everything else being equal, a right-wing think tank will tend to promote right-wing policy solutions, and a left-wing think tank left-wing ones.

Nevertheless, it would be premature to claim that *hysteresis* can explain the behaviour of think tanks before doing some leg-work. Even if one understandably presumes that think tanks are more likely to sustain the views they have held in the past than to argue for something entirely new, this is an industry where people and funding can move rapidly. A research team on a specific policy area can be swiftly formed, disbanded, or restructured, radically changing its policy focus and strategy, especially if the think tank in question depends to a great degree on project commissioning. Indeed, many think tanks fail to fashion a recognisable identity and research agenda, especially if driven too forcefully by disjointed short-term donor interests. Furthermore, this instability can be further compounded by political and economic crises. Still, *hysteresis* provides a useful model to frame the public interventions one is likely to find—a provisory hypothesis of sorts for intellectual dynamics.

One could go further, adding that a productive way of understanding think tanks is in their function as intellectuals, for, after all, their institutional continuity hinges on being perceived as 'legitimate' and helping create an environment where their ideas are viewed as such (Scott 1995). This implies that not only do think tanks garner and employ capitals but also participate in the process of defining them—for instance, by disseminating their views on what type of economic knowledge is valuable. That is why theories of intellectual life can be fruitfully employed to these organisations, especially to understand how, through their public interventions, an intellectual and political position—which implies views on others' ideas and on society at large—is asserted and promoted. The next section expands on these themes.

Think Tanks as Intellectuals

The business of most think tanks is to affect policy and the public debate, and for that their statements need to be seen as legitimate and worthy of attention. For that reason, their role can be compared to that of public intellectuals—or more precisely, 'collectives'[5] or teams of them. Discounting definitions of the intellectual that excludes those driven by utilitarian motives, treating think-tankers as such helps decipher important aspects of what they do. There are crucial similarities in the channels they intervene through (e.g., print and broadcast media, books), the resources they employ to be seen as 'in the know' (e.g., qualifications, language, even demeanour), and their place among different publics and networks (e.g., political contacts, benefactors).

[5]Bourdieu (1996 [1992]: 340) famously uses the expression 'collective intellectual' to mean those who draw upon and condense the knowledge of various independent actors situated across several fields.

Although I draw insights from several sources from the sociology on the topic, I concentrate heavily on Gil Eyal's (Eyal and Buchholz 2010; Eyal and Levy 2013) and Patrick Baert's (2012, 2015) writings. From the former, I take a focus on intellectuals' public interventions rather than on their social or normative characteristics. By 'public intervention' Eyal understands any act that communicates and mobilises knowledge, through any medium, in order to 'intervene' on issues of public concern. This expands the types of actors that can be construed as intellectuals and their possible modes of public engagement beyond the archetype of a Sartre or a Chomsky. Baert supplements this insight with close attention to the positioning and performative effects of public interventions; that is, how they situate those who utter them among other actors, amounting thus not merely to descriptions of the world but also to performative, illocutionary 'acts' in their own right that can change the world they describe. From that angle, position means both 'place' and 'vantage-point.'

Baert's 'positioning' theory is predicated on the existence of agents (in our case, think-tankers) who intervene through a medium (policy reports), a context where this takes place (the British polity, the 2008 crisis), and others who, in turn, *position* the initial intervenor (funders, policymakers, journalists, politicians, other think-tankers). Additionally, for the case of think tanks, being attentive to 'labels' and 'team membership' is useful. The first concept underlines the importance of signifiers that convey and package a set of assumptions about an individual or organisation—e.g., 'libertarian,' but also a think tank's very name, brand, and logo. The second refers to the fact that the production of think tanks' public interventions is a collective endeavour. They rely on many individuals with various skill sets to attract funding, write reports, promote them in the media, and enhance their impact, and hence some coordination is required.

A crucial advantage of these theories is that they clarify why intellectual change is, while improbable, not impossible. This is, first of all, because the task of becoming a recognisable actor is cumbersome and lengthy. A think tank might have to publish several reports and organise many events before it is seen as an important contributor on a given topic. Furthermore, intellectual change can foster demands for justification. If too frequent, the organisation could lose credibility and arouse suspicion among its supporters, which could threaten its very survival. Hence why the timing, context, and antecedents of public interventions matter. Changes of mind and heart never occur in a vacuum.

These ideas have two further implications. Firstly, for every intellectual team or coalition, what amounts to a relevant shift or realignment varies. For instance, think tanks positioned on the right, as a rule, tend to object to government action in the economy, but particular policy junctures or changing public attitudes might affect the intensity of this opposition. This space for internal tension and flexibility, between what is a matter of principle and what is an accommodation to particular circumstances, may have different thresholds and concern different policy areas for particular organisations. In other words, the place of the line that separates the 'sacred' from the 'profane'—between changes of mind that do or do not imply a challenge to a collective's core tenets—is an empirical matter (see Baert and Morgan 2015). Secondly, without the presence of significant external factors, there is no major incentive to change one's position—though the 2008 crisis, to be sure, provides an intriguing case of that.

There is, however, a crucial proviso to be made when applying these insights to think tanks. Most sociologists of intellectuals have focused on individuals rather than organisations, an important difference being that latter do not speak for themselves but need actual people to do so on their behalf. Think tanks are placed in the public debate through the public interventions of their actual members and, in turn, individuals think-tankers come to be associated with the 'brand' of their institution. In this sense, differences between the public interventions of various members of a think tank might betray internal tensions and be a sign of impending change. That is why, throughout this book, internal discrepancies are underlined wherever they are detected. To expand further on the organisational dimension of think tanks, in the next section I cover an interesting literature that has sprung on the role of ideas—and of those who advocate them—on organisational change and the policy debate.

Think Tanks as Institutions

Another tradition that informs this book is Neoinstitutionalism, especially in relation to how it theorises change in organisations and public policy. This body of scholarship emphasises the importance of the history and context of institutions to explain their behaviour, its main focus being on firms and the state (Guy Peters 2012). Its key principle is that the order in which events occur matters: those happening earlier are more likely to shape the trajectory of an institution, forming 'path-dependency lock-ins' (Pierson 2004). Given the weight of their history, and barring desperate circumstances, organisations tend to be reluctant to change once a certain stability and legitimacy have been achieved. For example, a think tank of a particular ideological persuasion might hesitate to modify its core arguments once a position in networks of support has been attained and reinforced over years of work.

On the subject of public policy, researchers indebted to Neoinstitutionalism have coined the notions of 'epistemic community' and 'policy paradigm' (Haas 1989; Hall 1993). The former denotes networks of experts who share notions of what is valid knowledge in a given policy area, policing who can speak authoritatively on it. The latter refers to the underlying assumptions that undergird policymaking. Think tanks have often been a focal point to those studying policy paradigms: two famous examples of these are indeed the 'welfare state' and 'neoliberalism.' Epistemic communities and policy paradigms are associated, in turn, with 'thought collectives' (Fleck 1979 [1935]; Dean 2012), social networks that seek to disseminate a way of thinking as widely as possible.

From this angle, intellectual positions tend to coincide with social and institutional realities, which has important implications for this book. In think tanks, the intellectual and the institutional tends to coalesce, crystallise into a recognisable position or 'brand.' A sustained belief in the importance of pushing a specific policy paradigm has major effects on the funding an organisation is likely to obtain, the policy areas it concentrates on, the staff it might attract, and the channels through which it reaches its audience.

However, Medvetz (2012) points out, Neoinstitutionalist scholars often mistakenly assume that the ulterior goal of think tanks is always policy impact. He has argued convincingly that, given their hybrid character, think tanks may well pursue other objectives, such as academic credibility, money, or the media spotlight. Furthermore, although much of Neoinstitutionalism rightly emphasises the importance of history, this focus can risk producing too stiff an image of organisations, one where change is unlikely and generally exogenous.

To confront that second objection, Colin Hay (2011) provides a necessary refinement. Much of this literature centres on the initial, formative stages of institutions rather than on later adaptations. Hay seeks to expand this focus, based on the idea that institutions are not simply bound by their history but are also 'forms of knowing,' and as such can adapt to changing circumstances. Thus, it is possible to observe 'post-formative' processes where a shifting environment affects how institutions operate. For instance, think tanks can alter their communications strategy towards one where media visibility becomes ever more relevant for policy impact, or be forced to do so by external demands in a process of 'isomorphism'[6] (Rich 2011).

This 'ideational' turn in the study of institutions has profound implications for think tanks—being, as they are, so wedded to their ideological function. Ideas are their business. Compared to other policy actors, the majority of think tanks have little or no formal accountability or institutionally sanctioned authority, and in most cases have only modest direct political clout. Barring a few cases, their reports are rarely referenced explicitly by policymakers, and their standing is grounded mostly on whether they are deemed worthy of being heard. However, these limitations also allow think tanks to intervene at a distance from the practical effects of their recommendations. That formal unaccountability and their presumed ability to suggest policy proposals that base their legitimacy on 'ideas' rather than solely on 'interests' are important factors behind their proliferation (see Stone 1996).

[6]Isomorphism refers to the process through which institutions become ever more alike as a result of mimesis and external pressures to conform (DiMaggio and Powell 1991).

Yet, given think tanks' orientation towards the policy debate, they have to be attentive to external circumstances—which requires reflecting on their relationship vis-à-vis governments and parties. On that topic, research on policy impact and its many intricacies led to significant developments within policy and science and technology studies. In a nutshell, difficulties in tracing policy impact prompted a rejection of rationalistic models of the link between knowledge and government and of overly systematic methods of measuring policy influence. For that reason, most policy scholars these days think that cutting-edge (socio) scientific knowledge seldom coincides with what is politically apposite, which often leads to maladroit compromises or opportunistic selection (Jasanoff 1995).

In that same line, a branch of Neoinstitutionalist policy studies has tended to focus on the power of narratives (Fischer and Gottweis 2012). Indeed, public policies themselves imply a narrative of the errors of the past and the possibilities ahead. Diagnoses of the problems and solutions involved in policymaking often take the form of stories rather than of disjointed sets of evidence. One could go further and say most policy narratives cannot do without 'crises,' moments in which fateful decisions have to be taken (t'Hart and Tindall 2009).

At this point, three scholars more closely focused on the role of research and knowledge in policymaking, and their 'timing,' are worth mentioning. The first is John Kingdon (2003), who stresses that policy problems are not given but constructed. According to Kingdon, policy change is most likely in 'windows of opportunity' where three 'streams' coincide: (a) the identification of a problem that needs addressing (e.g., a media event); (b) political conditions and will (e.g., parliamentary seats); and (c) practical feasibility to implement a solution (e.g., state capacity). In this context, think-tankers could be considered what Kingdon calls policy entrepreneurs "coupling solutions to problems and […] coupling both problems and solutions to politics" (ibid.: 20). Thus understood, think tanks act as depositories and 'peddlers' of pet solutions in search of a problem and often only become relevant, if ever, far in the future.

The conditions behind the emergence of windows of opportunity are, to be sure, mostly external to organisations—the 2008 crisis being, however, a clear instance. It follows that timing is of the utmost importance for the effectiveness of think tanks. Abelson (2002)—inspired by Sabatier's (1999) theory of the policy process—claims think tanks are most relevant in moments where the contours of a policy problem are being defined, which is when politicians are more receptive to external expertise. This can be advantageous, for instance, for think tanks keenly aware of internal party debates or in tune with the direction of public opinion (see Chapter 6).

A second scholar worth introducing here is Christina Boswell (2008, 2009). She argues that the role of research in politics and policy is most often, rather than substantial or instrumental, symbolic. Politically apposite research most often serves the purpose of legitimising a policy agenda and substantiating its underlying preferences, rather than necessarily shaping it. Following that insight, the role played by think tanks is to give credence to policies rather than to inform them. Being that think tanks occupy a murky brokering position in between knowledge production and politics, Boswell is not alone in questioning the extent to which the link they establish between ideas and policy is transparent (Stone 2007; Kay et al. 2013). They can play a role in helping determine which ideas from the academic and scientific world are considered policy-relevant—and that is, surely, not to say that these 'curated' ideas necessarily suffuse a given policy agenda, but merely that they are politically felicitous.

The third author I wish to cover is Stella Ladi (2011). Inspired by Boswell, Schmidt (2008), and the tradition of discursive institutionalism, Ladi distinguishes think tanks' discourse as either 'coordinative' or 'communicative.' The first is aimed at elites and is often instrumental and specialised (e.g., policy reports, parliamentary hearings). The second is targeted to the broader public (e.g., an op-ed, a tweet). Ladi believes that in junctures where public opinion is difficult to ascertain, communicative discourse becomes important for increasing the support for a policy agenda and framing the public conversation. Think tanks certainly engage in both types of discourse. Their first goal (shaping policy) requires attentiveness to the political field and its vicissitudes, while the second (influencing the policy debate) compels them to have a notion of which ideas are likely to be well received in the open.

In terms of timing, in our case at least four junctures coincide. First, immediately after the 2008 financial crisis there is an urgent appeal to determine what exactly happened. Second, given this demand for explanations, policymakers and the media are more likely to listen to external experts. Thirdly, and in tension with the latter, a mistrust of expertise becomes more prevalent, given their failure to predict or prevent the economic crisis in the first place. Fourthly, as think tanks are 'actual' organisations, the crisis can also affect their institutional environment, for instance by opening or closing funding avenues. As a consequence, the worth of available resources and the boundaries between fields of contention can become acutely uncertain, and think tanks are likely to eagerly engage in communicative discourse. In such a juncture, the role of intellectuals and experts becomes both more important and more suspect. Gramsci once said that in crises "the old is dying and the new struggles to be born" (1999 [1971]: 556).

Nonetheless, even when supplied with a cogent and timely narrative, think tanks can find it difficult to assert their views and attain a recognisable position in the public imagination. The most obvious obstacles to this are the availability of financial means, staff, and time, but access to networks and symbolic power are at least as important. In the next section, I expand on how think tanks' public interventions can reveal changes in both ideas and institutional logics, which frame the methodology of this book going forward.

Public Interventions as the Intersection of Ideas and Organisations

The concept of 'intellectual' or 'public' intervention has its roots in the idea of 'performative utterance'—from speech act theory, opposed to a representationalist view of language (Austin 1961)—meaning any speech act not merely 'representing' the world but acting upon it (Baert 2012: 310). Public interventions, thus understood, have an illocutionary character: not only do they provide a picture of the world, but can have consequences over what they describe and over those who utter them. For think tanks, these 'utterances' most often take the form of oral or written texts—though one should also include indicators, statistics, and other non-verbal forms of communication (Eyal and Levy 2013). An important part of these texts, given the publicity the policy debate demands, are made available online and are produced in various formats and with different audiences in mind. To interpret these, I propose below a working model of how think tanks operate in the public domain.

Public interventions have at least three dimensions. The first is a subject matter, policy issue, event, or state of affairs they refer to. What is considered worthwhile enough to intervene on depends, to mention but a few factors, on a think tank's research agenda, what is salient in the public debate, official policy, the issues political parties are pressing forward, funding priorities, the interests of trustees and funders, and regulatory constraints. Thus, any public intervention made on behalf of a think tank has in principle, a certain space for manoeuvre and limits when deciding what to intervene 'about.' Additionally, as policy issues are situated in a larger context, public interventions amount to entering into a 'conversation' with many other actors, often more specialised ones. Over time, public interventions on a similar issue tend to become crystallised in a 'research programme' and internal divisions of labour (research teams, clusters, or units).

Certainly, this 'entering into a conversation' is sought from a particular angle and accounting for previous interventions on the matter, which inescapably means highlighting some aspects of an issue over others. Accordingly, the very choice of 'problem' when studying a complex phenomenon can reveal political orientations and organisational characteristics. For example, focusing on monetary, financial, or fiscal policy to trace the origins of the crisis can be an indication of the institutional resources available to a think tank (e.g., expertise on an area) and of which factors are deemed relevant. Yet, some problems will be particularly challenging to avoid, the 2008 crisis being once again a prime example, given its severity, complexity, and uncertain consequences.

The second dimension of a public intervention is the 'narrative' of events it puts forward. By this I mean their substantive content, which customarily entails a description of a state of affairs, a recounting of its causes, and proposed measures to address it. Through this account, a specific understanding of the issue at hand is advanced while others are ignored or minimised. One could say such narratives weave together a plot that collects disparate elements into an 'intelligible whole' (Ricoeur 1980: 171), linking description, conclusions, and recommendations, often in the form of victims, culprits, and heroes. Ultimately, all narratives—even those by the most respected epistemic authorities, as the social studies of science literature reports—leave out some aspects of the social world and "[direct audiences] down specific logical channels, while blocking off others" (Hilgartner 2000: 9).

I hasten to add, these narratives can vary in scope, ambitiousness, and innovativeness. Depending on their focus, they can situate think tanks within a specialised field in which few participate (e.g., community development finance) or a more contentious policy arena in which many vie for attention (e.g., economic policy). Think tanks can attempt to posit a novel way of disentangling a policy problem, argue over the role of government, or merely aim to help policymakers reach informed decisions. These facets of 'narratives' inform the positioning of those uttering them, revealing their knowledge claims and political stance.

As devising novel policy narratives and ideas is taxing, public interventions by the same organisation often cite or evoke their previous work. Continuous efforts to herald a message through different mediums, developing its various aspects, can give distinctiveness and visibility to a think tank's 'brand' on a particular issue. Repetition, ultimately, is paramount to disseminate an idea and to gain recognition as its champion (Baert 2012: 316–317). Further, citing others, especially those with greater epistemic authority, can buttress one's account of events, especially for think tanks that are seen as less established or partisan. This might explain why organisations that seek a reputation of neutral expertise (e.g., NIESR, IFS) are commonly referenced by those advancing more overtly political claims (e.g., ASI, NEF). In turn, repetition and the citing of others one agrees with helps constructing and bolstering a network of like-minded allies, granting 'density' to one's position. Such a process has been described extensively for the case of free-market think tanks, which, at a comparative disadvantage in academia, referenced systematically the work of akin thinkers to build and disseminate a seemingly more robust case against the post-war consensus (Cockett 1995; Medvetz 2012; Stahl 2016).

In time, public interventions by the same think tank converge into what McLennan (2004: 4) dubs 'favoured arguments'—what I refer to, throughout the following chapters, as 'tropes.' Given that think tanks comment on a variety of issues, they tend to develop a distinctive form of argumentation. This is the case both for organisations undertaking commissioned research—where the choice of policy problem hinges at least partially on available project-based funding, and hence work on an often-unplanned range of topics—and for those with core-funding—which presumably have greater control over their focus. However, some public interventions can lack an explicit narrative in the form of a text, though one could arguably find one in their (specific) use of numbers, statistics, and indicators (Eyal and Levy 2013). As shown in Chapter 5, those producing econometric data to inform policy often have less control over how their work is interpreted, if perhaps being able to inform the common basis of the economic debate.

The third facet of public interventions is their 'format,' the concrete means, fashion, and media by which think tanks convey their 'narrative' to their audiences. This format has several aspects, such as timing, language (technical, polemic), intended audience (policymakers, the wider public), medium (social media, blogs, books, reports) and reference to other actors (academics, politicians, activists, journalists). To be sure, many formats can derive from the same narrative, but some are more suitable than others to produce particular effects, especially in terms of the audiences they reach. This is where the distinction between 'coordinative' and 'communicative' discourse becomes relevant (Ladi 2011). An inclination for a particular format in time generates a preferred mode of public engagement, a recurrent way of presenting one's organisation, and by implication situating oneself in the policy debate—e.g., plain-speaking or technical, passionate or aloof.

After continuous work on a number of policy topics, promoting a certain narrative, and employing particular formats, a think tank attains a certain 'position,' earning an organisation a recognisable 'brand.' This brand—the product of several concrete acts of 'speaking on behalf of' the organisation done by many—can facilitate obtaining funding, attracting promising job candidates, getting media attention, and gaining access to political networks. In Bourdieusian terms, here the procurement of symbolic capital is visible, where an intellectual position comes to be recognised as legitimate or marginal. At this point, the success or failure of a public intervention can be gauged, understood as the capacity of positioning the think tank among others, offering institutional recognisability, access to resources, and, sometimes, perceived policy influence. However, it is worth being mindful that success can mean different things for different think tanks (Medvetz 2012). It can mean policy impact but also reaching a distinct position among a particular audience.

This position, in turn, has effects on which policy areas a think tank is more likely to concentrate on and the type of funding it can pursue. For instance, if an organisation is successful in presenting itself as academically prestigious, research grants (such as those from the ESRC) become a possible source. If another is seen as staunchly in favour of free markets, it might be more inclined to seek support from corporate donors or wealthy businesspeople. Moreover, following Medvetz, the ability to muster the latter (corporate donations) can hinder its capacity to obtain the former (academic grants). For think tanks, perhaps all successes come with drawbacks.

Here it is worth exploring a possible limit to this model. Especially since the advent of New Labour and the dominance of a 'what works' discourse to justify policy decisions (Denham and Garnett 2006), many British think tanks have sought to position themselves as 'pragmatic,' devising practical policy solutions from a purportedly non-ideological standpoint. It suffices to skim through a few think tank mission statements to notice the ubiquity of the commitment to empirical rigour and reasoned argument, even if this is often not far from declarations of ideological commitment.[7] An environment where this tendency is

[7]Civitas' mission statement is a good example of this double pledge to empirical rigour and a particular political ideology. Civitas commits itself "to discovering how best to strengthen democracy, uphold limited government, maintain personal freedom, achieve opportunity for all, and encourage free enterprise [while] facilitat[ing] informed public debate by providing accurate factual information on the social issues of the day, publishing informed comment and analysis [...] Civitas never takes a corporate view." See (accessed 10 April 2016) http://civitas.org.uk/about-us/.

ubiquitous would, presumably, deter many from establishing too strong a 'brand,' as becoming too predictable could be counterproductive. As McLennan put it "the challenge [...] for a think tank is to achieve a certain identity, a certain 'brand,' without being perceived [...] as being enslaved by an ideology" (op. cit.: 2).

Even though a 'pragmatic' ethos could, in itself, be considered a brand—on account of its recurrence and the fact that it is certainly not politically neutral—a context where such approach is valued favours those who present themselves as capable of changing their mind, lacking an overly coherent paradigm. Given the imprint utilitarianism and gradualism have had on the history of British political thought (e.g., Edmund Burke on the right, the Fabian Society on the left) it is likely that most organisations would seek to be seen, at least to some extent, as flexible and moderate. Furthermore, given its magnitude, some form of intellectual change is bound to occur after the 2008 crisis, even in think tanks with the firmest brands. The question is of what kind.

Ultimately, I advance this model to reverse-engineer, as it were, the different components of think tanks' public interventions. From this exercise, it is apparent that they are both causes and effects of institutional dynamics—they give evidence of organisational change and can engender a new 'brand,' which in turn shapes a think tank's future output. Possibly this model's chief advantage, from a methodological point of view, is that it allows incorporating various forms of data across time. It offers a framework to connect public interventions together, highlighting that each utterance is more than a stand-alone piece but part of an iterative process that cuts across many platforms and audiences, both displaying and producing change. Furthermore, focusing on public interventions, by necessity, highlights internal tensions: one area of a think tank might be advancing certain narratives, while others operate in a completely different fashion. The process of producing and consolidating a think tank's brand opens the space for discrepancies and incoherence (see González Hernando and Baert forthcoming). The next paragraphs expand in more practical terms on which public interventions were employed and how, as well as on the rationale behind the interviews I conducted.

Public Interventions as Data

For methodological reasons, one type of public intervention is given pre-eminence here: policy reports. These are documents of variable length (generally of between fifteen and fifty pages, though some reach over a hundred), often funded by commission and sometimes involving original research, authored by think tank staff or fellow travellers to advance a policy proposal or approach while aspiring to a minimum of rigour.[8] As think tanks benefit from disseminating their ideas, most of their policy reports are free to access online as pdf files.[9]

Besides availability, policy reports are given priority for the following reasons. First, given that they demand greater resources than most other forms of public intervention, policy reports showcase a greater effort to position the think tank than, say, a tweet. Unlike a tweet or, presumably, most blog posts and op-eds, reports are unlikely to be produced at short notice. Second, since they are institutionally more costly, policy reports reflect the institutional realities of a think tank. They are often explicit about their organisational basis: for instance, regarding staff and the existence of a thematic team or unit. Thirdly, and relatedly, because other forms of public intervention (videos, opinion pieces, etc.) are often linked to these reports and aim to promote their findings. Indeed, much research funding comes with an item dedicated to dissemination, involving press releases, public events, and the like. Fourthly, because of the above, reports provide a privileged vantage point to observe larger changes in the intellectual and institutional environment of think tanks—for instance, in the form of available funding to research specific policy areas. Fifthly, the focus on policy reports is justified because, even if their format can be ill-suited for engaging with large audiences, their production can grant the possibility of being perceived as having made a considered argument in the first place. As once relayed by an audience member in a think

[8]Some think tanks, such as Civitas, IEA, and PX, have set up a peer-review process. These can operate as 'stress-tests,' as it were, for the plausibility of their arguments (Tchilingirian 2015).

[9]However, there are cases where reports have been pulled out of think tanks' websites. For an interesting example, see references to *The Hijacking of British Islam* in Chapter 6.

tank communications event, reports, although seldom read from front to cover, "have to have been written" (WonkComms 2014).

Regardless, considering the many other ways in which a think tank can present its ideas, other formats are also covered. For our cases, four noteworthy formats of public intervention are media appearances, blogs, public events, and academic publications. Presence in broadcast and written media can be illustrative of the avenues that are open to an organisation and the publics it is likely to reach. For instance, appearing frequently in the *Financial Times* reveals something about a think tank's position in relation to the finance and business sector. Blogs are especially significant for think tanks that rely on outside fellows for their policy reports and for those that seek to be responsive to ongoing policy debates, offering an avenue to provide quick commentary on the issues of the day. Holding public events is important for all think tanks, but especially so for politically well-connected institutions, as through this medium they bring together actors from different fields, strengthen their networks, and enhance their visibility. Finally, some organisations produce periodical publications that, while often relying on outside authors, can reveal transformations in the positioning of a think tank—notably NIESR's *National Institute Economic Review* (see Chapter 5).

The pre-eminence of these formats, as well as how they vary across time and organisation, are explored on an ad hoc basis throughout the following chapters—partly to illustrate how think tanks change over time and partly to showcase how their own environment changes. In tandem, policy reports were examined with the objective of reconstructing key junctures that marked these think tanks' intellectual output. Given the pressures think tanks faced to intervene after the crisis, written documents are privileged index of institutional change. Such texts can provide a wealth of data, including policy focus, authors, prevailing narratives of events, formats, language and (sometimes) funding. To be sure, as Bowen (2009) claims, although documents are seldom produced in the interest of researchers, they are useful trackers of organisational change.

The volume of policy reports published by a think tank can be considerable. Larger organisations can consistently produce over sixty per year. Although these reports share many traits, they can have remarkable diversity within and between organisations—in terms of length,

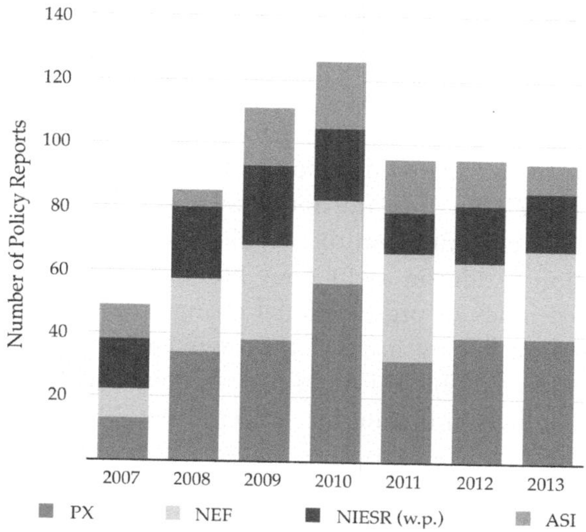

Fig. 2.1 Think tanks' policy reports per year (online) (As NIESR produces most of their public interventions on a rolling basis (journal articles, monthly forecasts) their working papers are listed here)

language, objective, audience, and policy area. For our cases, the number of reports produced by our four think tanks available online is 655.[10] The aggregated total is presented in the graph (Fig. 2.1).

These data already showcase some discernible patterns: the production of reports reached a noticeable peak in 2010. This increase was most notably bolstered by the 'Cameronian' Policy Exchange, and thus is conceivably linked to the advent of the Coalition government and its associated demand for policy ideas from their entourage. Nevertheless, the economic crisis itself elicited renewed interest in finance, macroeconomics, and cognate policy areas across all surveyed think tanks. From 2011 onwards, the total of published policy reports recedes and stabilises. These numbers also support the argument that think tanks are

[10]The fact that the reports hereby listed were those available—or at least referred to—online might skew the sample, under-representing the production of earlier years. However, reports published prior to 2007 are listed on the websites of all four think tanks, which mitigates that apprehension.

most active—and presumably, politically relevant—when policymakers are more attentive to external expertise, at the stage of the policy process when problems are being defined.

However, although public interventions in the form of text and other types of 'recordable communication' are useful, there are things they cannot do. Firstly, they do not provide much information about how they are produced, nor on the decision-making process that led to their publication (Atkinson and Coffey 2004). Secondly, although they are a privileged index of organisational change, they are precisely that, an 'index,' a trace left by internal processes which are hard to detect on paper (Bowen 2009). Thirdly, since they are 'final products,' they can obscure conflicts and minority voices (Boje 2001) and, as I show throughout the following chapters, internal tensions and incoherence are more common than frequently assumed.

The above is why I decided to complement documentary sources with interviews with members and former members of each case-study. My aim was to ask for a retrospective account of the history of their organisation, particularly with reference to the process leading to the publication of reports, in order to gather insights into how their work was planned internally. Through semi-structured interviews, I sought to 'weave together' different points in the history of each think tank, 'connecting the dots' between public interventions from the perspective of think-tankers themselves. This had the advantage of not betraying my position, allowing for the space for improvisation all elite interviews require (Dexter 2006; Harvey 2011). In total, I conducted interviews with seventeen members of the four think tanks, most in person but some over Skype, which lasted a minimum of an hour and a maximum of two. Current members of one of these think tanks (PX) did not reply to my requests, limiting my options to former employees. To avoid any conflict of interest, interviews were anonymised and, to make them even less identifiable to an informed observer, they are left unlabelled.

At this point, it is important to reflect on my own position as a researcher to detect its possible effects on the results I found, both in terms of how I am likely to be perceived by interviewees and in relation to my own biases. Some of my qualities might presumably, albeit unmeasurably, facilitate or hinder access and affect the data

I gather—male, young, non-British, white Latin American, then a doctoral student at the Department of Sociology of the University of Cambridge. While these characteristics could doubtlessly also affect my analysis, the effect of my political views and disciplinary background could be even more significant. Firstly, I define myself as broadly left-of-centre. So as to, at least partly, overcome any partiality that position may entail, I aimed to focus my analysis on intellectual and institutional change rather than on whether I believe a particular view of the economy and society is correct or mistaken. Secondly, I have no specialist training in economics. This at times meant I had to learn a new language, putting me in the position of a non-expert (Klamer et al. 1988: 77). There are pros and cons to this: it could avoid forcing the analysis into a 'straight-jacket' (Aberbach and Rockman 2002) while perhaps hindering my capacity to perceive significant yet subtle changes that someone with a stronger disciplinary background could have detected. However, it had one crucial advantage. During interviews, I interjected only when necessary to allow interviewees to develop their own narrative of events and views of what went on. In the process, I took intermittently the role of expert, mentioning names or reports when necessary to be seen as 'in the know,' and of the 'ignoramus' (McDowell 1998). This was facilitated by the fact that most think-tankers, even those whose work is almost purely technical, are eager to convey their ideas to non-specialists. Loosely based on the central premise of the strong programme for the sociology of knowledge—that not only the social conditions of 'false' beliefs should be studied, but also those of 'true' ones (Bloor 1991 [1976])—I looked at 'micro-histories' of policy-relevant knowledge production. The next and final section of this chapter explores how these micro-histories fit together and the model I employ to detect how they change.

The *Hysteresis* Hypothesis

As this book focuses on change, it is indispensable to be mindful of the timing of public interventions. To that effect, it is paramount to have a minimum awareness of key junctures, linking momentous economic and political events to discourses around them. As the studied period extends between 2007 and 2013, this comprehends some of the build-up of the financial crash, its outbreak, and its aftermath. In the British context, this includes heated discussions on the culpability of government and the City of London immediately after the crisis, the election of a Conservative-led Coalition in 2010 after thirteen years of centre-left New Labour rule, and the ensuing programme of public spending cuts. This setting, to be sure, varies across countries: in the United States, a stimulus package was implemented; in Greece and other Eurozone countries, austerity was seen as imposed from without rather than within; and many economies (such as those in Asia and South America) were affected to a much lesser degree by the crisis, at least in its initial stages. Certainly, not all think tanks will face the same pressures or refer to the same issues during those years, but an awareness of timing provides a frame and shared context that renders comparisons possible.

Of course, think tanks are not static; if anything, quite the contrary, as their craft demands keeping abreast of current events. Even the most intellectually rigid of organisations will vary in the policy issues their public interventions refer to, if not their arguments and formats, in reaction to their context, being on the 'defensive' or 'offensive' depending on where they see public policy moving. However, subtler yet stronger forms of intellectual change interest me more—those that could not be easily predicted by an informed observer. As explained earlier, *hysteresis*, the persistence of a view of the world even in the face of mounting challenges, is conceivably the most probable outcome after rapid transformations in an organisation's environment. This is perhaps particularly so for think tanks, as their hybrid character makes intellectual repositioning exclusively based on empirical evidence unlikely. Nevertheless, by the same token, their dependency on actors from other fields and their search of policy relevance can expose them to pressures to transform.

Employing *hysteresis* to frame the most probable interventions, I designed the following model (Fig. 2.2). Its aim is to illustrate the likely 'narratives' across time of those positioned differently in relation to the dominance of the financial industry and concerning academic and political networks.

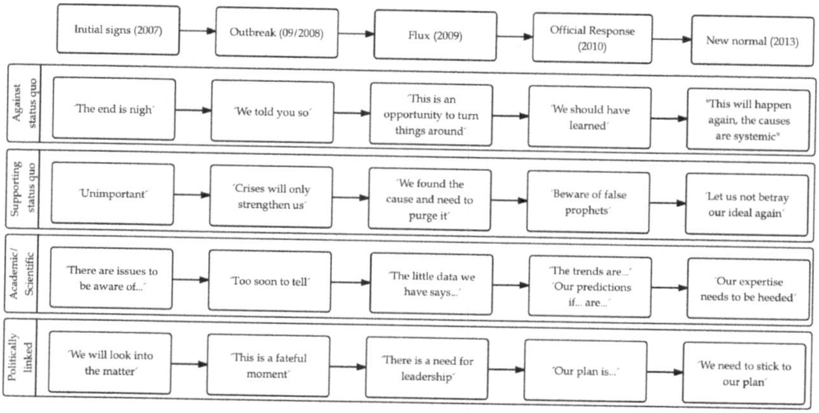

Fig. 2.2 The *hysteresis* hypothesis

In terms of timing, the crisis can be divided in:

- *Initial signs (2007)*: Indications of a looming financial crisis start to appear; actors might or not warn about its risks.
- *Outbreak (September–December 2008)*: Immediately after the fall of Lehman Brothers and the bailout packages in the United States and Europe, think tanks are impelled to provide an explanation to what is happening.
- *Flux (2009–2010 general election)* : A public and academic debate on the causes of the crisis and possible ways forward emerges. The public conversation is flooded with critiques of past policies and of mainstream economics.
- *Official response (2010–2011)*: After the 2010 General Election, fiscal consolidation became the main policy response to the crisis. Think tanks position themselves for or against the latter and evaluate its effectiveness.
- *New Normal (2012–2013)*: After a return to economic growth in the first quarter of 2013 and the entrenchment of the austerity discourse, new research priorities emerge and think tanks consolidate important institutional changes.

'Positioning' is the second dimension of this model, which is linked to how I designed the sample. Though this typology is not exhaustive and further cases could be added—e.g., close to an opposition party (IPPR); ideologically fickle (Demos), or even to supplement the positions I cover (e.g., IFS alongside NIESR, IEA alongside ASI)—it helps clarify the likely public interventions of our four case-studies across time:

- *Anti-status quo (NEF)*: For an organisation opposing financial deregulation and the primacy of free markets, the early phases of the crisis are marked by a vague sense of possibility, followed by disappointment and unease with the direction of policy, especially after 2010. They might also believe another crisis is likely, as the underlying causes of the first have not been addressed.
- *Pro status quo (ASI)*: For think tanks supportive of free markets and against most forms of financial regulation, the crisis represents, initially at least, a threat to capitalism. Later, however, the official policy agenda opened avenues for strengthening the free market by expunging it of distortions.
- *Academic/Scientific (NIESR)*: A policy institute with a socio-scientific ethos would start by assessing available evidence, seeking to inform the policy response to the crisis based on its expert authority. An arbitration of official policy would ensue, supported by specialised knowledge claims.
- *Politically linked (PX)*: In the case of a think tank connected to a party expected to become government, its goal would presumably be to produce proposals to inform a concrete political programme. A defence of said programme would likely follow.

This framework, it should be noted, represents only a conjecture limited to 'ideal types.' Were one to skim through their public interventions, it is highly probable one could easily find instances that fit each quadrant. However, that would commit the same intellectual offence many think tanks are accused of: selectively using empirical evidence to substantiate their preferences. I wish to avoid that risk by focusing more than on where this model applies, on where it does not. Through this approach, the occurrence or absence of *hysteresis* can be more readily appraised, and instances of intellectual repositioning can be highlighted. Through this design I seek to follow Viveiros de Castro's (2014: 13) research ethics: "[a]lways leave a way out for the people you are describing."

The promise of the *hysteresis* hypothesis, therefore, hinges on the theoretical yield of contrasting think tanks' likely and actual interventions. In the best-case scenario, this provides a framework for the tracing of intellectual and institutional transformations, enabling further comparisons across organisations and contexts. I judge that, by virtue of this framing, facilitated by a comparative case-study focus, this research can afford greater attention to detail and sensitivity to change than more ambitious projects with larger samples could.

However, there are two important limitations to this model. Firstly, the above hypothesis focuses mainly on the substantive content of public interventions. It says little about other forms of repositioning think tanks could undertake, particularly at the level of format. Indeed, as will become clear later in this book, given the orientation of these organisations towards the public debate, a changing media landscape, and the unstable status of economic expertise after 2008, some of the most important transformations think tanks underwent occurred at the level of how they engaged with their publics rather than on the substance of their arguments. Public interventions not only convey a view of the social world and the problems policy can solve, but also seek to build an implicit relationship between those who produce them and those they address.

Secondly, recollecting events across time to retell a story necessitates a degree of selection—deciding what is worth mentioning, what is part of the 'history' of a think tank. This is inescapable to any narrative of events, and as such, applies both to the work of think tanks themselves, to their self-presentation during interviews, and to this very research. Indeed, I cannot hope to reach a final 'truth' after retelling four versions of events—and this retelling is already an interpretation of interpretations (Geertz 1977), a narrative of narratives. The added value of this exercise rests on its comparative focus, on making its presuppositions explicit, and on its attempt to inform a reflexive sociology that takes heteronomous expertise seriously. I do not wish to claim the think tanks I study are correct or mistaken, but to examine how their tropes and views of the world are possible in the first place, and the mechanics of their change, if and when it happens. What I seek was, again, better expressed by Viveiros de Castro:

> [T]o take [natives] seriously does not mean to believe [...], to be in awe of what people tell you [...]. It means to learn *to be able to speak well* to the people you study [...] to speak about them to them in ways they do not find offensive or ridiculous. They do not need to agree with you *completely* — they will never do anyway; all we require is that they find our description a good enough one. It will always be a caricature of themselves, with certain traits exaggerated, others downplayed, certain points overstretched, others minimised. [...] As we know, oftentimes a proper, deliberate caricature captures the 'spirit' [...] of the person represented much more eloquently than a photograph. (Viveiros de Castro 2014: 17; emphasis in original).

Four empirical chapters follow, all structured in five sections. I start each by tracing the history of each think tank, showcasing relevant earlier work and the intellectual and institutional context from which it originates. The second, 'organisational and funding structure,' covers the organisational characteristics of each think tank, as well as its funding (depending on its degree of transparency). This is followed by a section entitled 'style and tropes,' where I explore some of the main themes behind the work of each think tank. A fourth weaves together a 'narrative of narratives,' interventions made during the years under consideration, and a final section draws general theoretical conclusions from each case-study.

References

Abelson, D. (2002). Do think tanks matter? Opportunities, constraints and incentives for think tanks in Canada and the United States. *Global Society, 14*(2), 213–236.

Aberbach, J., & Rockman, B. (2002). Conducting and coding elite interviews. *PS: Political Science & Politics, 35*(4), 673–676.

Atkinson, P., & Coffey, A. (2004). Analysing documentary realities. In D. Silverman (Ed.), *Qualitative research: Theory, method and practice* (pp. 56–75). London: Sage.

Austin, J. (1961). *How to do things with words*. Oxford: Clarendon.
Baert, P. (2012). Positioning theory and intellectual interventions. *Journal for the Theory of Social Behaviour, 42*(3), 304–324.
Baert, P. (2015). *The existentialist moment: The rise of Sartre as a public intellectual*. Cambridge: Polity.
Baert, P., & Morgan, M. (2015). *Conflict in the academy: A study in the sociology of intellectuals*. London: Palgrave Pivot.
Benson, R. (1999). Field theory in comparative context: A new paradigm for media studies. *Theory & Society, 28*(3), 463–498.
Bloor, D. (1991 [1976]). *Knowledge and social imagery*. Chicago: University of Chicago press.
Boje, D. (2001). *Narrative methods for organizational and communication research*. London: Sage.
Boswell, C. (2008). The political functions of expert knowledge: Knowledge and legitimation in European Union immigration policy. *Journal of European Public Policy, 15*(4), 471–488.
Boswell, C. (2009). *The political uses of expert knowledge: Immigration policy and social research*. Cambridge: Cambridge University Press.
Bourdieu, P. (1986). The forms of capital. In J. Richardson (Ed.), *Handbook of theory and research for the sociology of education* (pp. 241–258). New York: Greenwood.
Bourdieu, P. (1990). *Homo academicus*. Stanford: Stanford University Press.
Bourdieu, P. (1996 [1992]). *The rules of art: Genesis and structure of the literary field*. Stanford: Stanford University Press.
Bourdieu, P. (2000). *Les structures sociales de l'économie*. Paris: Éditions du Seuil.
Bourdieu, P., & Wacquant, L. (1992). *An invitation to reflexive sociology*. Chicago: University of Chicago Press.
Bowen, G. (2009). Document analysis as a qualitative research method. *Qualitative Research Journal, 9*(2), 27–40.
Calhoun, C. (1995). *Critical social theory: Culture, history and the challenge of difference*. Oxford: Blackwell.
Cockett, R. (1995). *Thinking the unthinkable: Think tanks and the economic counter-revolution 1931–1983*. London: HarperCollins.
Couldry, N. (2003). Media meta-capital: Extending the range of Bourdieu's field theory. *Theory & Society, 32*(5/6), 653–677.
Dean, M. (2012). Rethinking neoliberalism. *Journal of Sociology, 50*(2), 150–163.

Denham, A., & Garnett, M. (2006). What works? British think tanks and the end of ideology. *Political Quarterly, 77*(2), 156–165.
Dexter, L. (2006). *Elite and specialized interviewing*. Colchester: ECPR Press.
DiMaggio, P., & Powell, W. (Eds.). (1991). *The new institutionalism in organizational analysis*. Chicago: University of Chicago Press.
Dobry, M. (2009). *Sociologie des crises politiques*. Paris: Science Po, Les Presses.
Emirbayer, M., & Johnson, V. (2008). Bourdieu and organizational analysis. *Theory & Society, 37*, 1–44.
Eyal, G., & Buchholz, L. (2010). From the sociology of intellectuals to the sociology of interventions. *Annual Review of Sociology, 36*, 117–137.
Eyal, G., & Levy, M. (2013). Economic indicators as public interventions. In T. Mata & S. Medema (Eds.), *The economist as public intellectual* (pp. 220–253). London: Duke University Press.
Fischer, F., & Gottweis, H. (2012). *The argumentative turn revisited: Public policy as communicative practice*. London: Duke University Press.
Fleck, L. (1979 [1935]). *The genesis and development of a scientific fact*. Chicago: University of Chicago Press.
Gramsci. A. (1999 [1971]). *Selections from the prison notebooks*. London: Elecbooks.
Geertz, C. (1977). *The interpretation of cultures*. New York: Basic Books.
González Hernando, M., & Baert, P. (forthcoming). Collectives of intellectuals: Their cohesiveness, accountability, and who can speak on their behalf. *The Sociological Review*.
Gorski, P. (Ed.). (2013). *Bourdieu and historical analysis*. London: Duke University Press.
Guy Peters, B. (2012). *Institutional theory in political science*. London: Continuum.
Haas, E. (1989). Do regimes matter? Epistemic communities and evolving policies to control Mediterranean pollution. *International Organisation, 43*(3), 377–403.
Hall, P. A. (1993). Policy paradigms, social learning, and the state: The case of economic policymaking in Britain. *Comparative Politics, 25*(3), 275–296.
Harvey, W. (2011). Strategies for conducting elite interviews. *Qualitative Research, 11*(4), 431–441.
Hay, C. (2011). Ideas and the construction of interest. In D. Béland & R. Cox (Eds.), *Ideas and politics in social science research* (pp. 65–82). New York: Oxford University Press.
Hilgartner, S. (2000). *Science on stage: Expert advice as public drama*. Stanford: Stanford University Press.

Jasanoff, S. (1995). *Science at the bar: Law, science, and technology in America*. Cambridge, MA: Harvard University Press.
Kay, L., Smith, K., & Torres, J. (2013). Think tanks as research mediators? Case studies from public health. *Evidence and Policy, 59*(3), 371–390.
Kerr, R., & Robinson, S. (2009). The hysteresis effect as creative adaptation of the habitus: Dissent and transition to the 'corporate' in post-Soviet Ukraine. *Organization, 16*(6), 829–853.
Kingdon, J. (2003). *Agendas, alternatives and public policies*. New York: Longman.
Klamer, A., McCloskey, R., & Solow, R. (Eds.). (1988). *The consequences of economic rhetoric*. Cambridge: Cambridge University Press.
Ladi, S. (2011). Think tanks, discursive institutionalism and policy change. In G. Papanagnou (Ed.), *Social science and policy challenges: Democracy, values and capacities*. Paris: UNESCO.
McDowell, L. (1998). Elites in the City of London: Some methodological considerations. *Environment and Planning A, 30*(12), 2133–2146.
McLennan, G. (2004). *Dynamics of transformative ideas in contemporary public discourse, 2002–2003*. Accessed 15 October 2013. http://www.esds.ac.uk/doc/5312/mrdoc/pdf/q5312uguide.pdf.
Medvetz, T. (2012). *Think tanks in America*. Chicago: University of Chicago Press.
Medvetz, T. (2015). Field theory and organisational power: Four modes of influence among policy 'think tanks'. In M. Hilgers & E. Mangez (Eds.), *Bourdieu's theory of social fields*. London: Routledge.
Muniesa, F. (2014). *The provoked economy: Economic reality and the performative turn*. London: Routledge.
Pels, D. (1995). Knowledge politics and anti-politics: Toward a critical appraisal of Bourdieu's concept of intellectual autonomy. *Theory & Society, 24*(1), 79–104.
Pierson, P. (2004). *Politics in time: History, institutions and social analysis*. Princeton: Princeton University Press.
Rich, A. (2011). Ideas, expertise and think tanks. In D. Béland & R. Cox (Eds.), *Ideas and politics in social science research* (pp. 191–208). New York: Oxford University Press.
Ricoeur, P. (1980). Narrating time. *Critical Inquiry, 7*(1), 169–190.
Sabatier, P. (Ed.). (1999). *Theories of the policy process*. Boulder: Westview Press.
Savage, M., Warde, A., & Devine, F. (2005). Capitals, assets, and resources: Some critical issues. *British Journal of Sociology, 56*(1), 31–47.

Schmidt, V. (2008). Discursive institutionalism: The explanatory power of ideas and discourse. *Political Science, 11*(1), 303–322.

Scott, R. (1995). *Institutions and organizations*. Thousand Oaks: Sage.

Stahl, J. (2016). *Right moves: The conservative think tank in American political culture since 1945*. Chapel Hill: University of North Carolina Press.

Stone, D. (1996). *Capturing the political imagination: Think tanks and the policy process*. London: Frank Cass.

Stone, D. (2007). Recycling bins, garbage cans or think tanks? Three myths regarding policy analysis institutes. *Public Administration, 85*(2), 259–278.

Swartz, D. (2013). *Symbolic power, politics and intellectuals: The political sociology of Pierre Bourdieu*. Chicago: University of Chicago Press.

t'Hart, P., & Tindall, K. (Eds.). (2009). *Framing the global economic downturn: Crisis rhetoric and the politics of recessions*. Sydney: ANU Press.

Tchilingirian, J. (2015). *British think tanks and the production of policy knowledge: A social network analysis of policy intellectuals* (PhD thesis). Department of Sociology, University of Cambridge.

Viveiros de Castro, E. (2014). *Who is afraid of the ontological wolf? Some comments on an ongoing anthropological debate*. CUSAS Annual Marilyn Strathern Lecture, Cambridge. Accessed 20 November 2015. https://www.academia.edu/12865685/Who_is_afraid_of_the_ontological_wolf.

Weaver, R. K. (1989). The changing world of think tanks. *PS: Political Science and Politics, 22*(3), 563–578.

WonkComms. (2014). *Taking the pulse: The role for evidence in the election debate*. Accessed 30 October 2015. https://www.youtube.com/watch?v=wI2Sv8mD-3k.

3

The New Economics Foundation: Crisis as a Missed Opportunity

The New Economics Foundation (NEF) is a centre-left think tank which aims to offer innovative policy proposals with a focus on the environment, wellbeing, and alternatives to free-market economics. It was founded by the leaders of 'The Other Economic Summit,' which ran parallel to the 1984 London G7 meeting. NEF's launch is historically associated with a growing conviction within some policy circles that social development should be understood more broadly than merely as measurable economic output (Friedmann 1992). Hence the need for a 'new economics,' which the book *The Living Economy*—edited by the first head of NEF in the year it was established—defines as "based on personal development and social justice, the satisfaction of the whole range of human needs, sustainable use of resources and conservation of the environment" (Ekins 1986: xiii). One could also mention the influence of an earlier book by the Club of Rome, *Limits to Growth* (Meadows et al. 1972), which predicted that, if

In this chapter, as in those to follow, citations in the format (MM/YYYY) refer to the corresponding think tank's policy reports and blog posts, which are listed separately in the bibliography. Also, some of the data in this chapter and in Chapter 4 inform González Hernando (2018).

M. González Hernando, *British Think Tanks After the 2008 Global Financial Crisis*, Palgrave Studies in Science, Knowledge and Policy, https://doi.org/10.1007/978-3-030-20370-2_3

current trends continue (overpopulation, pollution, resource depletion), there would be a collapse of the global economy within a century. These ideas NEF opposes to what is considered a self-defeating 'old' economics, which "boils down to the pursuit of economic growth, as indicated by an increasing Gross National Product [under the] assumption [...] that growth is good and more is better" (ibid.: 5–6).

At first glance, NEF seems to follow loosely the model of Richard Bronk's (2009) 'romantic' economists. They champion a view of economics that distances itself from the formalistic methods of the neoclassical paradigm that has dominated said discipline for decades. That necessitates an understanding of economic agents that diverts from the abstractness of the *homo economicus (*oriented by a rational and narrow pursuit of profit and self-interest), acknowledging the role emotions, culture, history, institutions, communities, and the environment play and should play in economic and policy decision-making. However, that is a negative definition, delimiting what NEF is not rather than what it is, and is broad enough to encompass Marxists, Keynesians, Malthusians, Polanyians, Minskyans, and environmental economists. Indeed, there is ample debate on the internal unity of heterodox economics, some even claiming it should be considered a separate discipline from its mainstream, supply-side counterpart—effectively, a *new* economics, so to speak (Cronin 2010). For the purposes of this book, it suffices to point out that this ambiguity grants NEF a space to manoeuvre that more traditional intellectual positions could not afford. It allows for a diverse if not necessarily systematic array of intellectual resources from which to draw upon.

Nevertheless, if NEF has one main intellectual inspiration, it most certainly is E. F. Schumacher (1911–1977), author of *Small Is Beautiful: A Study of Economics as If People Mattered* (1973).[1] Initially under Keynes' tutelage—of whom he distanced himself later in life—he is among the most influential post-war heterodox economists of the twentieth century, and advised the governments of Britain, India, Germany, Zambia, and Burma. Although Schumacher never proposed a

[1]NEF's motto, 'Economics as if people and the planet mattered' signals both their indebtedness to Schumacher and their focus on the environment.

comprehensive model of the economy, he offers a set of values and priorities to guide economic and social development—most noticeably in essays such as *Buddhist economics* (ibid.). Succinctly put, Schumacher's ideas constitute a plea for an economy that "put[s] human wellbeing at the centre of economic decision-making and everything within the context of environmental sustainability" (Schumacher 2011: 10). If any citation can encapsulate NEF's self-understanding, it is probably the latter.

Since the 1990s under the management of Ed Mayo (1992–2003)[2] and Stewart Wallis[3] (2003–2015)—and, beyond our timeframe, Marc Stears (2015–2017) and Miatta Fahnbulleh (2017–present)—NEF garnered growing visibility, earning the title of *Prospect* 'think tank of the year' in 2002. It was an important partner of 'Jubilee 2000'—an international campaign for the elimination of the sovereign debt of third world countries—and produced several impactful research streams. Among these one can highlight *Clone Town Britain,* which assessed the damage wrought by chain stores on local businesses and the economic diversity of local high streets, as well as publications on macroeconomics (Huber and Robertson 06/2000), the environment (Muttitt 02/2003), and wellbeing (the Happy Planet Index, HPI, since 2006).

Organisational and Funding Structure

Around the time when interviews were conducted for this research, NEF's main office was located in an unassuming—discounting the emblazoned logo over the entrance—red brick building in a small residential road near Vauxhall station, a few blocks away from where its new offices are. This is not far removed from the more think tank dense Westminster-Millbank area but is markedly separated from it by the Thames. Inside the premises, a two-storey open-space office made for a somewhat teeming and buzzing workplace, if lacking the infrastructure to host large public events.

[2]Ed Mayo later became general secretary of Co-operatives UK.

[3]Stewart Wallis OBE is also member of the Club of Rome, Vice-Chair for the World Economic Forum's Global Agenda Council on Values, and former International Director of Oxfam.

In the British context, NEF is a relatively large think tank, variably of about 50 employees, publishing an enviable number of reports. In the period between January 2007 and December 2013, NEF released over 174 reports. Being NEF a 'generalist' organisation and a self-proclaimed 'think-and-do tank,' its staff produces research on several policy areas, structured thematically in semi-autonomous research programmes or teams. They comprise—for the period under consideration and after several processes of restructuring—'Finance and Business,' 'The Great Transition Initiative,' 'Natural Economies,' 'Social Policy,' 'Valuing What Matters' and 'Wellbeing.'[4] These have relative independence in their funding sources, outreach capacities, and connections with outside experts and practitioners. Indeed, research commissions and grants are often targeted to specific teams rather than to the whole organisation. In line with this structure, most interviewees claim NEF's leadership—or at least Stewart Wallis—has a non-interventionist approach to management, granting teams the space to develop their own research agenda. The work carried out by these includes:

- *Finance and business*: banking reform; local banking[5]; community currencies;[6] community development finance; industrial strategy.
- *The great transition* initiative: local economies; Great Transition model; New Economy Organisers Network (NEON).[7]
- *Natural economies*: climate change; energy; fishing; international development; transport and infrastructure.

[4]These have included in the past 'Centre for the future economy,' 'Local economies' ('Connected economies' before 2010), 'Climate change and energy,' and 'Democracy and participation.' Most of these programmes have been rebranded, (temporarily) suspended, or merged with the aforementioned research teams.

[5]The Happy Planet Index (HPI) ranks countries according to life expectancy, measures of wellbeing, and their ecological footprint.

[6]NEF has advocated for some time the use of local currencies to strengthen local economies and businesses, producing research and advocacy on the matter. For instance, NEF members of staff were involved in the setting up of the *Brixton pound*.

[7]The Great Transition model seeks to devise and implement modes of social and economic organisation that deliver wellbeing within sustainability constraints. Meanwhile, the New Economy Organisers Network (NEON, est. 2013) attempts to bring together actors such as NGOs, charities, trade unions, and other organisations to develop pathways towards a 'new economy.'

- *Social policy*: time banks; shorter working week; local effects of austerity; migration; welfare reform.
- *Valuing what matters*: measuring inequality; Social Return on Investment (SROI).[8]
- *Wellbeing*: measuring wellbeing; wellbeing for policy; HPI.

From this non-comprehensive list, the scope of NEF's work is already apparent; even without considering publications by research teams that have been discontinued or restructured. Many of these areas seem, at first glance, *sui generis* and somehow unconnected from each other. This heterogeneity is partly due to NEF's decentralised and flexible funding structure, but it might also be related, as shall be shown below, to a malleable albeit holistic view of its overall mission.

NEF also has links with a number of organisations with similar aims, be it through the dual membership of some of their researchers and fellows (e.g., Ann Pettifor, NEF fellow and head of PRIME), connections from former employees, and direct involvement of particular research teams—e.g., Time Banks UK was set up with the support of NEF's Social policy team. Being NEF a charity, to these we could add the links implicit in the composition of its board of trustees, which includes members associated with other think tanks (Simon Retallack,[9] Howard Reed[10]), consulting companies (Leo Johnson,[11] Jules Peck[12]), charities and NGOs (Sam Clarke,[13]

[8]'Social Return on Investment' is a methodology for measuring the social, economic, and environmental impact of charities, businesses, and public services, which is employed by several NEF research teams.

[9]Simon Retallack is Associate Director for Policy and Markets at the Carbon Trust and former Associate Director at the IPPR. Data on this trustee and those below were retrieved from NEF's website unless otherwise stated.

[10]Howard Reed is the Director of Landman Economics and former Chief Economist at the IFS.

[11]Leo Johnson is a Partner in PwC's Sustainability and Climate Change team.

[12]Jules Peck is strategic advisor at Edelman and Founding Partner of Abundancy.

[13]Sam Clarke is Chairman at the Low Carbon Hub, former Chairman for Friends of the Earth UK and former Fundraising director for Oxfam.

Colin Nee,[14] Lyndall Stein,[15] Sue Gillie[16]), and, beyond our timeframe, academia (David McCoy,[17] Jeremy Till[18]). Internationally, NEF has links with the New Economy Coalition in the United States, a fast-growing think tank with similar values.[19]

Of all left-of-centre think tanks in Britain, a minority to be sure, Demos is possibly its closest likeness, which is perceived by NEF staff as less ideologically coherent and more tightly driven by management. Comparable institutions, such as IPPR, Compass, and the Resolution Foundation, can be distinguished by their close links with the Labour Party, their funding structure—membership organisations, substantial core donations—and NEF's focus on the environment and sustainability, even in areas not commonly associated with environmental concerns.

Between 2007 and 2013, NEF has had an overall positive if unstable financial situation, bordering a total budget of £3m. As a requisite for its charitable status, NEF publishes annual statements that in their case display the sums, sources, and types of funds they receive (Table 3.1). NEF's restricted income, defined in their accounts as that "to be used for specified purposes as laid down by the donor," constitute around half of its revenue. This proportion is important because the ratio between restricted and unrestricted funds impacts the type of public interventions NEF is more likely to produce. It affects NEF's responsiveness in the short term: whether it can swiftly appoint people and resources to intervene in a particularly salient policy area or react to unforeseen events. Likewise, in the long run, NEF's funding structure

[14]Colin Nee is Chief Executive of the British Geriatrics Society and former Interim Executive Director of Reprieve.

[15]Lyndall Stein is former executive director of Concern UK.

[16]Sue Gillie is Chair of the Trustees at Clean Conscience and former trustee at the RSA, Paxton Green Timebank, Network for Social Change and Oxford Research Group.

[17]David McCoy is a senior clinical lecturer in global health at Queen Mary University and Director of the global health charity Medact.

[18]Professor Jeremy Till is Head of Central Saint Martins and Pro Vice-Chancellor, University of the Arts London.

[19]Formerly New Economics Institute (which NEF helped set up in 2009), and before that the E. F. Schumacher Society (founded in 1980).

Table 3.1 NEF financial overview[a]

NEF funding (£)	2007	2008	2009	2010	2011	2012	2013
Restricted income	1,733,861	1,339,715	1,115,613	1,081,756	991,048	1,919,127	1,490,051
Unrestricted	1,576,365	1,315,923	1,921,404	1,615,475	1,514,020	1,366,934	1,626,236
Total income	3,310,226	2,655,638	3,037,017	2,697,231	2,505,068	3,286,061	3,116,287
% Unrestricted	47.6	49.6	63.3	59.9	60.4	41.6	52.2
Total expenses	3,046,891	2,700,811	2,673,052	2,994,314	2,652,117	2,772,911	3,263,330
Balance	263,335	−45,173	363,965	−297,083	−147,049	513,150	−147,043

aData from financial statements supplied to the UK Government Charity Commission (2008–2013, Reg. No. 1055254), accessed 22 October 2014, http://apps.charitycommission.gov.uk/Showcharity/RegisterOfCharities/FinancialHistory.aspx?RegisteredCharityNumber=1055254&SubsidiaryNumber=0

affects its capacity to direct its research agenda to areas where it does not coincide with donors' funding preferences. That is, in a nutshell, whether it will concentrate most of its energy in producing 50-page reports on the economics of takeaway food (Sharpe 06/2010) or in areas deemed a priority.

Restricted funding is mostly project-based, targeted at particular programmes or for the writing of specific reports, and is generally supplied by charitable trusts and governmental bodies—which range from local development agencies to the Scottish Office of the First Minister and the European Commission. In time, each research team tends to generate links with specific funders, whose interests are often geared towards precise policy issues. What these donors seek from reports, for the most part, can be divided into capacity-building—e.g., producing best practice guides for charities (NEF 05/2009)—research—e.g., studying the local level effects of austerity on benefits claimants (Penny 07/2012)—and advocacy—e.g., raising awareness of the importance of the environment for wellbeing (Esteban 10/2012).

Some of the most significant donors of restricted funds, in terms of both volume and length of engagement, include the Hadley Trust, AIM Foundation, and the Barrow Cadbury Trust, all charitable trusts. By way of illustration, the Hadley Trust funded work (Brown and Nissan 06/2007; Nissan and Thiel 09/2008) on community development finance institutions (CDFIs) providing micro-lending for disadvantaged communities and the 'Valuing What Matters' team. The Barrow Cadbury Trust has supported research, not only for NEF but also for IPPR and Demos, on the consequences of austerity in local services provision and immigration. And Ian Marks CBE, head of AIM Foundation, was a prominent backer of Jubilee 2000, and his trust supports NEF's Wellbeing team.

The pre-eminence of these types of sources raises two questions. The first concerns the aims of charitable trusts in funding think tanks in general and NEF in particular. Although this goes beyond this research's purview, a few things can be said on the matter. These trusts are often set up by wealthy individuals or organisations, and their aims are generally philanthropic. For instance, the Hadley Trust's stated mission is to "assist in creating opportunities for people who are disadvantaged as a result of environmental, educational or economic circumstances."[20] To achieve this, the Hadley Trust commits part of its endowment to two think tanks, NEF and Policy Exchange (PX). The fact that it funds policy institutes on opposite sides of the political spectrum—as does Barrow Cadbury Trust and many others—suggests that political inclination is not *tout à fait* an excluding factor in these donations.

Nevertheless, there are limits to what can be financed in this manner. First, because what charitable trusts fund is somehow aligned with institutional objectives, however broadly understood. This can mean a prioritisation of narrow policy areas or specific types of output (policy reports as opposed to books, for instance). These priorities can affect what a generalist think tank that veers for their support, such as NEF, occupies its time on, which in the long term can drive institutional change. One interviewee noted that after the 2008 crisis, many researchers working on environmental issues moved to other policy areas, partly because "funders were seeing climate change as something less urgent" (NEF interview).[21]

The second question that arises from the prominence of charitable trust funding in a think tank's finances concerns its influence in the latter's intellectual output. As described by interviewees, research contracts customarily involve a series of meetings and emails where each party negotiates the objectives and execution of a report. This gives space to policy experts to expand on their arguments without much intrusion, barring that the policy area itself is already decided. Funders might highlight one aspect of a report's findings over others when it is promoted, but the research design, results, and recommendations are most

[20]Accessed 24 October 2014, http://apps.charitycommission.gov.uk/Showcharity/RegisterOfCharities/FinancialHistory.aspx?RegisteredCharityNumber=1064823&SubsidiaryNumber=0.

[21]Something similar can be said of funding from government bodies. This is not (only) because they are unlikely to commission research that might question current policy, but since, in a climate of austerity, public funding for research is scant and its focus ever narrower. For instance, in 2010 the Local Economies team had to be restructured, as it "had a lot of funding from regional development agencies [which] the Tories scrapped [...] in their first week of government" (NEF interview).

often left to think-tankers themselves. Indeed, offers of funding are sometimes rejected, and attempts at editorial interference are rare and frowned upon. As a think tank's accrued reputation for rigour and advocacy effectiveness allows access to this type of funds in the first place, excessive and noticeable involvement of funders in the research process is to be avoided. Thus, a report's publication requires the conciliation of the funder's motives for supporting research and the image, style, and research agenda of the think tank concerned.

It should be noted, nonetheless, that there are restrictions, however tacit, to the content of what a charitable trust funded report can argue. Funding for systemic or theoretical research on the economic system is rare, if rising (NEF personal communications 06/02/2015). In the long term, without other funding sources, this permeates to an important part of a think tank's output, affecting its position *vis-à-vis* other policy actors. Closing the circle, this position in turn affects the funding avenues a think tank such as NEF can pursue more favourably. In time, this business model could risk skewing the output NEF generates, in terms of research topic (driven by funder's interests), format (mainly 40-page reports), and content (circumscribed, and unlikely to be a frontal attack on the economic system). One interviewee said, evocatively, that this funding arrangement "creates a treadmill" which can make the enterprise seem like a "pdf-producing machine" (NEF interview).

The rest of NEF's endowment is made up of its unrestricted funds. These are particularly sought after, as they allow the organisation for greater discretionary spending and to be less constrained by donors' interests. In terms of the provenance of these funds, an important proportion originates from the charitable activities of research teams, which coincides with their grants and donations from research funding bodies. Another relevant source is the income generated by NEF-Consulting, a trading subsidiary that offers consulting services to businesses, public bodies, and charities guided by NEF's core principles, "enabling organisations to move towards a new economy."[22] In practice, this means the

[22]See (accessed 18 October 2014) http://www.nef-consulting.co.uk/about-us/what-we-do/.

application of itsSROI methodology—devised by the Valuing What-Matters team—to small businesses, public sector organisations, and charities. It was set up both to garner further funds and to demonstrate the applicability of NEF's model, and although initially operating at a loss, it has grown steadily. By the end of our timeframe, it produced a profit that grants NEF a small but growing space for manoeuvre.[23] Going forward, and considering the institutional capacity built around NEF-Consulting, it is conceivable that the SROI methodology could help expand the reach of NEF's ideas across different fields. For instance, it could allow for NEF's insertion in infrastructures of expertise on the link between economics and wellbeing. In that sense, the SROI, HPI, and other indicators produced by NEF could be construed as interventions in their own right (Eyal and Levy 2013) that could situate NEF at the infrastructure of economic policymaking linked to wellbeing.

Another part of NEF funds between 2007–2013 came from unrestricted donations. These can be divided in those made by trusts—especially AIM Foundation, Freshfield Foundation (since 2010), RH Southern Trust and Roddick Foundation (until 2011)—and those made by individuals, including fundraising campaigns,[24] which most years make up roughly a tenth of NEF's income. Finally, revenues from the sale of publications and educational products (such as the New Economics Summer School) make a minor addition to the final budget.

In terms of the forms of funding NEF does not receive—which can be just as telling as those that it does—one can mention large individual and business donors (with a notable exception in 2009),[25]

[23]According to NEF's financial statements, in 2009 NEF-consulting cost the organisation around £86,000, while in 2012 it made a profit of roughly the same amount. In 2017, NEF-consulting profits after overheads and royalty payments were similar, though its trading volume increased considerably.

[24]Sometimes fundraising can connect to communication initiatives in quite interesting ways. See: 'just bond' campaign, accessed 22 September 2014, http://www.neweconomics.org/press/entry/just-bond-offers-return-of-a-brighter-future-for-the-planet.

[25]According to NEF financial statements, in 2009 an anonymous donor made a one-off contribution of £500,000 in unrestricted funds. This has not happened since, and most other years the total of received individual donations rarely exceeds £200,000.

political parties, academic research councils, and core donations from Government, membership programmes, and subscriptions. The reasons behind the absence of these sources are sometimes related to idiosyncrasies of the British context—by way of contrast, Germany provides public funds for think tanks—while others are specific to NEF. For instance, according to interviewees, NEF's distance from neoclassical economics (still dominant in academia), mainstream political parties, and pro-corporate advocacy hinders its prospects of accruing support from, respectively, economic research councils, party donations, and big businesses.

These figures also attest that just one bid or donation can have a considerable effect on NEF's overall finances. Sometimes, as in 2009 or 2012,[26] a single source can skew the funding structure towards restricted or unrestricted income, and towards operating at a profit or at a loss. This might explain the frequency of team restructuring. Regardless, its relative flexibility and internal diversity allow NEF to 'hedge' the risks inherent to an unstable organisational model. The 2008 crisis thus, for NEF, was not only a chance to disseminate its message, but also presented organisational risks (e.g., less available public funding) and opportunities (e.g., banking reform research commissioning).

Style and Tropes

Since the policy areas generalist think tanks focus on are likely to be swayed by available funding—especially for those without substantial core funds—listing them is not enough to outline an organisation's work and identity. What is more, in order for think tanks to be effective, their intellectual 'brand' must be translatable to multiple domains. Much like new-right think tanks in the 1970s applied neoliberal

[26] In 2012 the Tubney Trust—set up by the founders of Blackwell publishers—granted NEF £400,000 in designated funds upfront for a three-year project to "develop the capacity of key environmental NGOs to understand and utilise socioeconomic data in their work to achieve biodiversity benefit in UK waters," accessed 22 October 2014, http://www.tubney.org.uk/annual%20report_2012.pdf.

recipes across hitherto uncrossed policy boundaries (Mirowski 2013), NEF needs to generate its own variety of advice, along with a cogent and coherent way of presenting its case across publics. This applies to democracy, higher education, and immigration, where NEF has produced relatively few reports, as well as to banking reform, social policy and wellbeing, where it has published many. For those reasons I focus on think tanks' 'tropes,' recurrent ways they present an argument and engage with audiences, rather than only on their policy focus and explicit recommendations.

A qualification is necessary here. Focusing on tropes and themes might grant an artificial homogeneity to the publications of a think tank, or at least predispose us to find such unity, especially for complex and diverse organisations such as NEF. Moreover, tropes and formats vary not only across authors and research teams but crucially, over time. With these caveats in mind, what follows is a summary of NEF's recurring themes and tropes to position it in the context of other think tanks and policy actors. I contend, however, that part of what follows applies somewhat less to some of NEF's interventions after 2011–2012—with the Coalitions' spending cuts programme under full swing—and especially since 2013, after the report *Framing the economy* (Afoko and Vockins 08/2013) was published.

One striking characteristic of much of NEF's discourse is a recurrent dualism that helps distinguish what is desirable from what is not. These exist in some form in many other think tanks and generally coincide with intellectual positions that precede organisations themselves. In NEF's case, the sides of the dichotomy are drawn around the axis of a formalistic mainstream economics—aimed at consumption and growth for their own sake—versus a new economics, focused on wellbeing and sustainability. This is expanded further by oppositions like infinite growth/planetary limits, greed/fairness, clone towns/transition towns,[27] 'too big to fail' banks/local banking and 'business as usual'/new economy.

[27]These refer, respectively, to the homogeneity of high streets across the UK and to a movement, with which NEF has links, aimed at building resilient and sustainable local communities (Kjell et al. 06/2005).

Recurrently, NEF authors position themselves as outsiders, acknowledging they are in a minority in economics and policymaking circles, or at least seldom on the side of the status quo. They are, in that sense, an *avant-garde* pushing for a new relationship between the economy, society, and the environment. In consideration of this position, the resolution of the aforementioned dualities, and the window of opportunity for positive change, often involves both an apocalyptic and an utopian dimension: beginning with exhortations of how wrong things are, and how worse they will become if their warnings are unheeded, many of their reports set out a vision of how society can improve, not least because it will be forced to. As these radical ideas respond to fundamental and ultimately unavoidable policy choices, NEF's attempts to engage with unevenly receptive publics and policymakers depend on raising awareness of the seriousness of the situation and of our responsibility to act. Hence, crises—be they visible or abstract, current or looming—are key to their argument, as they constitute turning points in the development of their narrative. This is well summarised in a quote from NEF director Stewart Wallis: "[a] different future is not just necessary, it is also possible" (NEF 10/2009: 2).

A good illustration of the latter is the exercise of policy science fiction *Future news* (Boyle et al. 05/2009), where NEF authors illustrate their vision of the future through fictional tabloid pages set in 2027. In them, the threats of peak oil and climate change are taken much more seriously than at present, as they are more 'actual.' Dystopian events fill *Future news*' pages, including East Anglian climate refugees, food shortages, floods, blackouts, and demographic crises. As a response, coming British governments institute a carbon credit policy (a cap on the carbon footprint each individual is allowed to produce), subsidise wind farms, promote the consumption of local produce, as well as other proposals from prior NEF reports.

Another way of expressing this opposition between hope and looming crises is through the re-assessment what is already present. NEF documents often argue that run-of-the-mill economic analysis rarely values what is not or cannot be priced (e.g., carbon footprint, wellbeing), and hence that important aspects of human existence habitually go unnoticed, cast aside as 'externalities.' This argument is connected to the work of the 'Valuing What Matters' team and its SROI methodology, binding together policy areas such as work, local economies, environment, and wellbeing. This set of themes, in turn, merge into public interventions and policy proposals that amount to a reappraisal of our current situation through socio-scientific methods—in the case of SROI and HPI, by linking economic outputs with environmental and wellbeing indicators. In such a way, NEF seeks to bring to the fore what is foundational yet often overlooked. The role of the state and government in this juncture is generally, if implicitly, that of a facilitator of change and a regulator that accounts for 'externalities' neglected by the market (Green New Deal Group 07/2008), another point of contention between NEF and free-market think tanks.

One of the most far-reaching of their policy recommendations, linking avant-garde thinking and the everyday, is the *21-hour week*, funded originally by the Hadley Trust. A star project of the social policy team, it proposes to drastically shorten Britain's working week, or at least move towards such a policy goal in line with other European nations. This measure, they contend, would have favourable effects on productivity, employment, gender equality, wellbeing, and the environment. It has been advocated for years through numerous avenues, including policy reports (Coote et al. 02/2010), public talks (LSE 2012), books (Skidelsky et al. 2013), and the media (*Guardian* 2010).

This example illustrates another interesting feature of NEF's work. Although the arguments behind *21-hour week* are mostly developed by the Social Policy team, its diagnosis and recommendations are also conceptually linked to other parts of the organisation: to 'Wellbeing' and 'Valuing What Matters' concerning the importance of free time for wellbeing, and to the 'Environment' and 'Local Economies' concerning their positive externalities. Thus, while there is a high degree of compartmentalisation in NEF, the recommendations they produce frequently adjoin several policy issues: the economy, society, wellbeing and the environment. In this sense, NEF's work tends to be malleable while holistic, which perhaps explains why it is sometimes seen as a think tank that joins dots across policy areas, despite a funding structure not amenable to cross-thematic work. One interviewee said:

> [t]here are strengths and weaknesses to [NEF's] disaggregated system. One of the key weaknesses is that it becomes very difficult to do cross-thematic work. Ironically really, because that's one of the things people tend to think NEF is very good at […] but actually the reality is, and this is probably because of the funders and the way that the funding system works, we don't do nearly as much of that integrated work as we'd like. (NEF interview)

In terms of the format and tone of NEF's publications, a few rhetorical devices are particularly common throughout their reports and public interventions, which allow staff to generate a recognisable 'brand' across platforms. One of them is the use of quotes from renowned intellectuals, including sociologists (Bauman, Beck, Giddens), writers (Wilde, Thoreau), philosophers (Mills, Bentham, Aristotle), politicians (Churchill, Gandhi), and heterodox economists (Keynes, Schumacher). Many of their public talks and policy briefs are accompanied with quotations from such figures, often with anti-consumerist undertones and critiques, veiled or otherwise, of mainstream economics. This, one can infer, both associates them with those thinkers and showcases a degree of symbolic and cultural capital, in Bourdieusian parlance. A good example, by heterodox economist Kenneth Boulding—in NEF's *Great Transition* (10/2009: 4)—is this quote: "Anyone who believes exponential growth can go on forever in a finite world is either a madman or an economist."

Another common rhetorical device is to cite international examples. These tend to be divided in two: European countries with strong welfare state provision (e.g., the Netherlands concerning its four-day week and high productivity) or Latin American countries (plus Bhutan): Uruguay for the frugality of (former) President Mujica; Costa Rica for their low carbon footprint, high levels of wellbeing, and high performance on NEF's HPI; Bolivia for their Mother Earth laws; Bhutan for the focus of its policies on happiness. These are compared positively against the UK, the implication being that other policy agendas are both possible and practical.

These two types of argumentation are linked to NEF staff's self-perception of the organisation's heterodoxy and its often-unconventional policy recommendations. In the words of interviewees, NEF "doesn't do dogma" and is preoccupied with "making a radical case, even if it doesn't seem feasible in the short term" (NEF interview). As NEF's vision is in a minority—especially in a public debate mostly dictated by austerity—it seems necessary to demonstrate that their position is both intellectually robust and a possible guideline for the future: NEF's ideas are already being applied elsewhere and have the support of great thinkers.

It is worth adding that, in NEF's conjunction of the 'necessary,' the 'possible,' and the 'desirable,' there is a curious bringing together of normative and expert-descriptive discourse. Like any think tank, NEF needs both to provide a vision of what a polity should strive for and support its claims with evidence. Considering its position as an outsider, the latter is fundamental, but being sceptical of academic economics, it must rely on ideas that are outside the mainstream. This conundrum begs the question of whether NEF should follow a strategy of seeking to debunk economics by recourse to its very methods or instead distance itself from them. One could say that, barred from many institutionally sanctioned sources of legitimacy for their economic arguments—in the form of Economic and Social Research Council (ESRC) funding, for example—it must to a degree redefine expertise, and fashion its own form of being technocratic. Arguably this is the prime mission of the Valuing What Matters team and its indicators.

That tension between advocacy and expertise helps explain another latent tension, mentioned by some of NEF staff, between becoming "a green version of the IFS" (NEF interview), or a more politically engaged organisation: the dilemma of how to be both *hacks* and *wonks* (Medvetz 2012: 173–174), present in some form in all think tanks. Nonetheless, for most intents and purposes, the strain between these two strategies is not resolved for the totality of NEF's interventions but varies across research teams. Maybe this is part of what grants think tanks such as NEF their flexibility, often in conflict with their unity.

This conundrum is also at the base of the institution of the 'fellow.' NEF fellows are individuals appointed by management, who are linked to the organisation through their work or ideas. Although not involved in day-to-day operations, fellows collaborate by co-authoring reports, writing blogs, or speaking in public events citing their affiliation. Among them we find former employees (Andrew Simms, Anne

Pettifor, Daniel Boyle, Nic Marks, Ed Mayo), scholars (Edgar Cahn, Ian Gough, Tim Jackson), and policy experts and practitioners (Jeremy Nicolls, David Woodward). They undertake, mostly independently, a significant amount of interventions across a broad range of policy areas, generally from an academic or semi-academic standpoint, often invoking their 'fellowship' status. Some examples of these, from particularly prolific fellows, are David Boyle's (2014) work on the decline of the middle classes, Anne Pettifor's (2003, 2006) books on credit-fuelled debt, Andrew Simms' (2013) book on causes for optimism, and Tim Jackson's (2011) *Prosperity Without Growth*. Fellows thus help build and disseminate the institution's brand across platforms, if being less directly coordinated with the rest of their work.

A Frenzy of Activity: The Crisis as Seen by Critics

As some of NEF's central ideas are that GDP growth cannot be extended ad infinitum, that modern economics does not measure nor foster wellbeing, and that the neoliberal model is unsustainable, this think tank arguably had a terrific opportunity to disseminate its message after 2008. As early as 2000, they had criticised the British economy's over-reliance on financial services (Huber and Robertson 06/2000). Plus, fellow Anne Pettifor (2003, 2006) can claim to have anticipated an impending crisis in the US credit market. Thus, as a guiding framework for the public interventions it was likely to produce, I hypothesised in the previous chapter that NEF's position would first involve reading the situation as a confirmation of their exhortations against unfettered markets, and later they would plead for radical changes in the economic model. In the British context, given that austerity became the main policy response to the crisis after 2010, this would be followed by disappointment and critiques of government policy. This section, and analogous ones in the chapters that follow, seek to trace this process, gauging whether such a sequence took place and, perhaps more crucially, how.

NEF, given its longstanding criticisms of financial capitalism, can be said to have had a 'footing' (Harré et al. 2009) to explain the causes of the crisis and denounce the excesses of finance to a public that was then more likely to be receptive. This realisation produced, in the words of one interviewee "a frenzy of activity, and a sense of elation" (NEF interview). NEF initially capitalised this opportunity firstly by compiling old material and reports that had become relevant again. As the proverbial 'garbage can' (Stone 2007) NEF sought to promote its ideas by gathering work that had been prepared for decades and draw attention to earlier reports that now seemed eerily prescient (Johnson et al. 10/2007). Being NEF outside the status quo, the juncture elicited, more than a corporate defence, a sustained effort to disseminate its critique of free markets.

A first public intervention came in the form of a panel organised with *The Guardian* (2008a), later collected in the report *Triple Crunch* (Potts 11/2008). In it, several speakers from the Green New Deal Group (see below) argued for the necessity of a new economic model, now this one had been proved wanting. This was followed by two presentations at the Leeds Schumacher Lectures[28] in October 2008, delivered by Andrew Simms and Anne Pettifor, later published as *Nine Meals from Anarchy* (Simms 11/2008). Soon after, a more fleshed out report, *From the Ashes of the Crash* (Simms 11/2008) was released, with an extensive list of policy proposals. Among these one finds, quite early on, some which became popular among critics of the City: splitting the retail and investment banking, stronger regulation of the financial industry, capital controls, and closer oversight over the production of money. Simms, then Policy Director, had launched in August 2008 *One hundred months* (*Guardian* 2008b)—a series of columns in that present a countdown until global temperatures rise beyond 2 °C, a point of no return according to many climate scientists. In public interventions such as these, crises in the economy, society and the environment became one and the same. Baert (2012: 315) has stated that repetition is often necessary for a successful act of intellectual positioning; at the beginning of the crisis, this was certainly the case for NEF.

[28]These are lectures, associated with the Schumacher College, where much of NEF staff has taught.

A multitude of similar undertakings was carried out across research teams, and this recounting will necessarily be incomplete. Nonetheless, a few generalisations can be made. First, understandably, in this juncture, much of the tone became faintly 'legislative' in Osborne's (2004) terms: both denunciatory and normative, condemning the mistakes of those who brought the global financial system to the brink of collapse and expressing how they conceive things should be. An example of the former is the report *I.O.U.K.* (Boyle et al. 03/2009), which accused 'too big to fail' institutions for their inability and unwillingness to lend, arguing for a descaling of the financial industry. In this juncture, one also finds reports that call for radicalism and optimism. The Social Policy team, for instance, published *Green Well Fair* (Coote and Franklin 02/2009), setting out a new framework to orient social policy towards fostering wellbeing and sustainability. In tandem, NEF campaigned to promote Timebanks in the UK—organisations in which people trade in time instead of money—through *The New Wealth of Time* (Coote et al. 11/2008). This brought to the fore older work on the topic (NEF 07/2001) and sought, in a period of economic uncertainty, to foster a response to the crisis that strengthened local communities and created spaces outside the monetised economy. This initiative has since then been mirrored across recession-hit countries such as Spain. Appositely, around that time, Stewart Wallis spoke at the World Economic Forum meeting in Davos (Reuters 2009).

Roughly during the same period, NEF published two important documents that need to be underscored: some time before the crisis, *The Green New Deal* (07/2008) and soon later *The Great Transition* (10/2009), in reference to Franklin Delano Roosevelt and Karl Polanyi, respectively. The first marks the foundation of the Green New DealGroup (GND). Composed of some of NEF's senior members of staff, two former directors of 'Friends of the Earth,' the economics editor of *The Guardian*, Caroline Lucas (Green Party MP), the former head of Greenpeace's Economics Unit, tax justice campaigner Richard Murphy, and the Chairman of Solarcentury (a solar energy company), it is an attempt to bring together a proposal to strengthen the economy through public investment aimed at fostering green energy and sustainable development. Based on the idea of an imminent 'triple crunch' of energy (peak oil), environmental (global warming) and economic crises, the GND proposes a route to tackle all three in an environmentally friendly Keynes-inspired stimulus plan. It could be considered an attempt to reconcile the 'Schumacherian' and 'Keynesian' strands within NEF's thought.

The Great Transition, on the other hand, was drafted initially as a bid for support from the funding body 'Network for Social Change.'[29] It amassed much of the work NEF had been producing over the years, but also assembled in a coherent vision, the teams and expertise available in the organisation. The transition referred, in this sense, to a movement in economics and society, from consumption to wellbeing, from greed to the common good, from derivatives to the 'real' economy, and from the abstract to the local.

Their names alone display these papers' ambitiousness, and they exemplify NEF's bridging of the apocalyptic and the utopian through a critical juncture—a transition, a new deal. Both are, explicitly, a roadmap to influence future governments on what the next steps should be to tackle current predicaments, at a holistic level and in consideration of the dire consequences where it otherwise. In that sense, while arguably at a greater distance from politics and parties, these documents are faintly reminiscent—while made from a position of lesser political capital—of the large policy blueprints other think tanks have produced for oncoming governments; such as the Adam Smith Institute's 'Omega project' in the 1980s (Pirie 2012a: 93–105) and PX's reports in 2009–2010 (see Chapter 6). But even if unable to exert a direct influence over policymaking or parties' election manifestos, these reports serve as a declaration of principles and as a platform for organising their direction of travel. They collect, as it were, the ideas and proposals NEF stands for and set a roadmap for the future of their organisation.

To launch their Great Transition, NEF organised the event *The Bigger Picture* (see Wimbush 02/2010) held in October 2009 in London's South Bank, with over 2100 reported attendees. Although relatively late in the development of the crisis, the aim of this was to bring together funders,

---

[29]Accessed 25 October 2014, http://www.thenetworkforsocialchange.org.uk/uploads/docs/NSC%20funding%20report%202011%20with%20addendum.pdf.

campaigners and practitioners in order to articulate new forms of social development. Funding for such a venture was uncertain, and arrangements took longer than expected. One interviewee told me NEF did not manage to raise the required resources for holding it, but it was deemed necessary to continue, even if at a loss. Such was the urgency of bringing like-minded people together to develop a new progressive economic project.

More generally, at the time there was a resurgence of Keynesian arguments, and a window of opportunity seemed open to influence future governments. Austerity, as a policy agenda, was still a looming possibility, but other options were deemed conceivable. The then Prime Minister Gordon Brown even argued explicitly for the necessity of a GND (BBC 2009)—not to mention that, beyond the timeframe of this book, the concept has regained salience, especially in the United States. This possible rise to centrality, some interviewees point out, produced hopefulness and restlessness, but also disagreements within the organisation, especially concerning the role NEF should take: were they to be activists or technocrats for a new order?

A good example of how these tensions played out in practice concerns the report *A bit rich* (12/2009), based on the SROI methodology. It consisted in an attempt of pricing of the social return of different professions, compared to their average salaries. It found, for instance, that City bankers destroy £7 for each pound they receive, while hospital cleaners produce over £10 for each they are paid. But to reach these figures, its model passed to the bankers the full responsibility for the losses of the banking crisis, which prompted very public fault-findings from the *Financial Times* (2009) and others. Although the report succeeded in terms of receiving ample coverage, the methodology was often criticised both within and outside the organisation.

It is worth remembering that the years after the crisis also saw the vertiginous rise of the internet and social media as political tools. Possibly inspired by a successful RSA series, NEF commissioned animated videos to be promoted through YouTube. One of the most remarkable was the Impossible Hamster,[30] which argued against infinite

[30]Accessed 25 October 2014, https://www.youtube.com/watch?v=Sqwd_u6HkMo.

economic growth by picturing that a hamster, growing indefinitely at the same rate it does between birth and puberty, at the end of a year would weigh nine billion tons, so why assume the economy can grow forever at an exponential rate? (Chowla et al. 01/2010). This piece, while highly popular and drawing from years of work in an amusing fashion, had an evident difficulty: is a recession the moment to argue we need less, not more, growth? But when if not then? This came to embody NEF's tension between an orientation towards sustainability and the need to produce an alternative to austerity, which most often means some form of economic stimulus.

Be that as it may, the main event defining NEF's public interventions within our timeframe was, to be sure, the election in 2010 of a Conservative-led government. It came to crystallise the primary policy response to the crisis: fiscal consolidation through cutting government spending, especially on welfare. The window for structural transformations to our economic system began to close, and a GND started to sound unlikely. Like much of the left, NEF gradually found itself opposing the tide of change instead of leading it. As early as 2009, the GND Group had argued that "cutting spending now will make the recession worse by increasing unemployment, reducing the tax received, and limiting government funding available to kick-start a Green New Deal while there is still time" (12/2009: 3).

In terms of NEF's organisational structure, 2010 saw research teams such as 'Climate and Energy,' 'Local Economies,' and 'Democracy and Participation' suffering from persistent funding shortages, which forced restructuring. 'Finance and Business' and 'Social Policy,' on the other hand, received substantial boosts, particularly after a winning a long-term funding bid from the European Commission on banking reform, and support from charitable trusts to study the local effects of austerity (see Penny 07/2012; Penny and Slay 08/2013). The Finance team saw a growth in staff numbers and counted from 2010 onwards with a group of talented economists with practical experience in banking (Tony Greenham, Lydia Prieg). They released a series of ambitious reports tackling tax havens and the role of the UK's government in subverting financial reform (Greenham et al. 03/2011), and directly advised the Independent Commission on Banking (ICB). At the time, the organisation also hired a former Treasury advisor, James Meadway, as Senior Economist. Meadway was tasked with developing an analytical model to respond to the recession and austerity and wrote frequently for NEF's blog. On this greater focus on macroeconomics and finance, one interviewee commented:

> It's definitely the case that, after the financial crisis [...] funders were prepared to fund work looking and financial system reform, whereas before their eyes definitely would have glazed over. (NEF interview)

Nevertheless, as NEF's income structure tends to be disaggregated and that researchers often act as fundraisers for their teams, this meant skewing NEF's research towards topics where funds were available, leaving fewer people and less time available to respond to oncoming events. Succinctly put, since NEF relies heavily on research project grants, much of their resources had to be dedicated to securing and carrying them forward, rather than to more ambitious longer-term work, or even to reacting to a volatile policy environment. Presumably, this is linked to the political positioning of the institute, which, some interviewees claim, make it difficult to gather substantial core funding:

> If you've got six months ahead of you until you run out of funding, the programme will cease. It's very precarious. You have to keep getting these grants [...] It can be very hard for us to have capacity 'there,' ready to respond to stuff that happens [...] because who would pay for that? It's very difficult to go to a funder and say 'can you just employ me for a year and, you know, I'll just do stuff?'. (NEF interview)

Concerning NEF's policy impact, even though the Coalition's policies were often diametrically opposed to their recommendations, there were still coincidences and attempts to bridge differences. A good example is NEF's efforts to reframe Cameron's Big Society programme (see Chapter 6) to accommodate its ideas, such as through the co-production model—noticeably in the report *Cutting it: The big society and the new austerity* (Coote 10/2010)—while warning about the perils of rapidly descaling social service provision. Nonetheless, the impact of their ideas was scant in mainstream political circles, with some notable exceptions. Three we can mention are the receptiveness of the government to measure wellbeing—which might position centrally NEF indicators such as the HPI and SROI—, their report to the ICB and, at the EU level, the role of the 'Natural economies' team in the debate over the European Common Fisheries Policy reform.

Also, partly as a response to what they saw as a lack of knowledge on how banks operate, NEF published a book (Greenham et al. 2012) on money creation that has since been cited by the Bank of England and *Financial Times* editor Martin Wolf (2014). In this line, NEF has also been involved in a growing movement within economic departments to expand the curriculum; it has a strong presence in *Rethinking Economics* events (Yang 06/2013). One could theorise that being part of initiatives to broaden economic analysis—which have been growing following 2008, such as George Soros' sponsored Institute for New Economic Thinking (INET)—is of great interest to NEF, at least at two levels. First symbolically, as it legitimises their intellectual position—redefining, in Bourdieusian terms, the boundaries of the field of academic economics—but also materially, as it expands the resources they could muster (e.g., academic funding, graduates interested in heterodox economics). Nonetheless, a fundamental shift away from the neoclassical paradigm is yet to materialise fully in much of the economics discipline.

Discounting these partial successes, a sense of frustration lingered among many of my interviewees. In late 2013, the GND Group published a five-year review, stating that: "[o]ur warnings and our advice [...] went unheeded" (09/2013: 8). For NEF, a particularly upsetting facet of austerity was the widespread impression that government overspending caused the crisis. It went against everything they had been arguing, especially on the need for a different macroeconomic strategy (Meadway 04/2013) and investing 'upstream' rather than cutting preventive initiatives, making matters worse in the long term (Coote 04/2012). And yet the discourse of austerity was often considered 'commonsensical,' visible in the speeches of government representatives, the media, other think tanks, and even in polling results. Unsurprisingly then, there was often a disconnect between some of the work produced by NEF on the one hand, and the political mood and policy agenda during the Coalition government, on the other. For instance, in late 2012, when public spending cuts and the associated austerity discourse were in full swing, NEF proposed a *National gardening leave* (Simms and Conisbee 10/2012), encouraging employers to offer days off for gardening as a way to produce locally, increase wellbeing, help the environment, and strengthen local communities.

To understand and overcome the pervasiveness of the austerity narrative became an important part of NEF's mission, as it was considered the main obstacle to pursuing the type of policies it advocates for. A first attempt was a series of reports and blog posts, *Mythbusters*, which sought to debunk common arguments for austerity by showing that they are factually inaccurate; hence their often-technical tone. For instance, against the myth *Britain is broke*, they argued:

> If Britain is broke at the moment, then [...] it was also broke for a whole century between 1750 and 1850, and for 20 years after the Second World War. In reality, in neither case did the UK default, and reveal itself as bust – both periods were times of investment and national renewal. Today, our national debt is significantly lower than Japan's (about 200% of GDP), and comparable to Germany's (83%) and the US (80%). By international or historical standards, the national debt is not high. (Reid 04/2013).

A few months after *Mythbusters*, a new publication appeared that took a different tack: *Framing the economy* (Afoko and Vockins 08/2013), written by two new employees with an interest in narrative analysis. This report sets out a diagnosis of why the austerity story has proved so resilient, as well as a toolbox to counter it. Influenced by the literature on semantic framing (Lakoff 2004), it explained that simply arguing against austerity without questioning its foundations can in fact reinforce its frame, the way of looking at the world that underpins it. This showcases a growing concern with the ineffectiveness of anti-austerity discourse and conveyed the realisation that "[t]he battle for the economic narrative will be won with stories, not statistics. Armed with facts alone, opponents of austerity stand little chance against a story that is well developed, well told and widely believed" (Afoko and Vockins, op. cit.: 30).

For instance, the frame 'dangerous debt' that the *Mythbusters* series tried to debunk, even if believed to be factually inaccurate, can conjure a persuasive way of understanding events:

> This frame existed long before the austerity story. Most of us already hold negative frames about debt, often rooted in personal experience. What the austerity story does is combine our existing fears about debt with our understanding of what is wrong with the economy. This means that single words are likely to activate the frame: borrowing, debt, deficit, credit, national debt. (Afoko and Vockins, op. cit.: 9)

What is more, the report argues that some of the arguments made by NEF in the past were not only ineffective, but sometimes even counterproductive:

> Spending cuts are never presented as desirable; their part in the austerity story hinges on the idea that there is no alternative to them. This is a very powerful way to frame an argument, suggesting there is no choice to be made. It is the lynchpin of the austerity story, the part that you must accept to make the plot believable. People who argue against austerity by stressing the pain it causes are not attacking this frame – depending on their language they may even be reinforcing it. (Afoko and Vockins, op. cit.: 9).

In another instance, *Framing the economy* even criticises the way ideas in many NEF reports are presented, given their highly technical style of argumentation and their inability to be linked to everyday understandings of economics, especially in the context of a public discourse steadily tilting to the right. Afoko and Vockins even go as far as to critique the environmentalist 'prosperity without growth' idea for its untimeliness in a period of recession. Nevertheless, intellectual changes are often difficult and slow to implement in all instances and across all of the organisation. One interviewee pointed out:

> If you're working on something like banking [...] I'll very much find myself [...] talking in terms of neoliberal frames, talking in terms of 'you should do this, you should have local banks, because of their effect on growth' [...] which obviously reinforces the focus on GDP, and it probably isn't helpful, but I find myself slipping into it [...]. Sometimes you find yourself reinforcing negative frames just because you're trying to make your point. (NEF interview)

After years of campaigning for ideas that had had relatively little effect on government policy, and after a momentous crisis that has been read by many in the exact opposite way NEF researchers understood it, a new strategy had to be taken. But to subvert the coalition's frame, NEF not only needs a rival story, but messengers to carry it forward persistently and coherently. Since 2013, Daniel Vockins has helped organise the New Economics Organisers Network (NEON), a networking platform to put in contact people (in advocacy organisations, charities, academia, public sector workers, etc.) interested in campaigning for a new economy, along with exploring, in post-2008 Britain, what 'new economy' means in the first place.

On Being 'Neffy'

To finish this chapter, I would like to highlight two particular public interventions by NEF researchers that epitomise the internal tension the organisation underwent between 2007–2013. Both depict widely different types of public engagement made by important figures associated with the organisation. One could even say they represent the antipodes of what NEF—and maybe even left-wing think tanks more broadly—can encompass and can be. The first is Nic Marks' (fellow and former head of Wellbeing) presentation of the 2010 HPI results for TED talks (TED 2010). The second is James Meadway (Senior Economist) addressing Occupy London Stock Exchange (LSX) protesters at the height of their movement in 2011 (Meadway 11/2011). Although both are short speeches, they are very different in most other respects. The first sets forth an optimistic message, arguing that policy should focus on what is important for wellbeing, guided by what international survey data suggests. Marks' speech was delivered in an uplifting tone, reminiscent of much of NEF's material circa 2009. The second is more pugnacious, a denunciation of an unfair and intolerable state of affairs. Meadway relies on official figures to chastise some of the City and government, calling for collective action against austerity and the dominance of finance in a manner that is rousing and unequivocally political.

The contrast between these public interventions is illustrative of two features of how a think tank like NEF operates. First, across its staff and teams, NEF is capable of mustering several kinds of resources, targeting them to different audiences in a way that no individual working on her own could. It is indeed difficult to imagine Meadway and Marks occupying each other's places, not adding to the mix, for instance, NEF head of Finance Tony Greenham advising the ICB in a much more formal context (NEF 12/2010).

The range of ways in which NEF staff can present themselves and their organisation cannot be explained only by the varying conventions implicit when working on many policy areas and political arenas. Different ways of conveying the same message, to be sure, affect how this message is read and those uttering it are perceived. Thus, through differences in the public interventions of its policy researchers, the tensions marring NEF's work—between being seen as activists or experts, generalists or specialists—become visible to attentive outsiders too. Baert and Booth's (2012) four dichotomies of how public intellectuals present themselves—preferring hierarchy or equality, generality or expertise, passion or distance, and presenting themselves as individuals or as representatives of collectives—are thus built inside the organisation and resolved each time someone speaks on NEF's behalf.

The second feature I wish to discuss concerning how NEF's members speak on behalf of their organisation is the fact that public interventions also vary over time. Firstly, in terms of their relation to external events. Being think tanks such heteronomous institutions, the timing and effectivity of a public intervention cannot be dissociated from, for instance, the official policy agenda and the focus of media organisations. The external context shapes how an idea is understood and positions the organisation: Meadway's speech in Occupy LSX could be read very differently in any other context.

With respect to NEF's capacity to react to changing external circumstances, although the organisation's financial structure requires it to concentrate most of its resources in mid-term commissioned projects, a few strategies have made it more responsive in the short term and innovative in the long term. These include linking project-based research to the wider debate and the organisation's core mission whenever possible, allocating restricted resources to dissemination, recuperating older reports in the 'garbage can' model, and assigning some time to focus on quick-response public interventions (for instance, through blogs and social media). Nevertheless, it is perhaps unavoidable for an organisation so dependent on research contracts that "most of the innovative ideas that [NEF has] come up with [...] have almost been done in people's spare time" (NEF interview), and that the best-timed public interventions are often done outside the 'treadmill.'

Furthermore, as much as public interventions by NEF are influenced by external circumstances, they are also refracted through their own institutional history. Making the argument, for example, that the UK needs to regulate more tightly its banking industry can have very different consequences for an organisation in 2006, 2008, or 2013. This variability is also illustrative of transformations in the organisation in terms of staff, structure, and funding, as well as at the level of how the institution 'thinks' itself in a turbulent political context.

Concerning this point, it is worth noticing that ideational change is very often associated with movements of staff. Cohorts of researchers entering or retiring can leave traces in the publications of an organisation, as well as in their thinking. Ideas of what NEF is and should be are, one could say, 'embodied' by actual people. Concerning the arrival of Daniel Vockins, one interviewee commented:

> [Vockins] has read a lot of books on narrative and the power of narrative and how, when you look at how things change, actually facts don't matter. What matters is narratives and stories. He brought that into NEF and he said 'look, we need to talk about this.' So, had he not been at NEF, I don't know if that conversation would have happened. (NEF interview)

These new conversations, nonetheless, rarely occur abruptly and never do so in a vacuum. Changes in think tanks are seldom total and radical; they are sedimentary and layered processes. By way of illustration, then NEF head Stewart Wallis, in a speech entitled *The flawed dominance of economics* at the University of Cambridge in late 2013, made no mention of framing, retaining much of the discourse typical of 2009–2010 (CUSPE 2013). In 2014, however, he penned an op-ed for *The Guardian* (2014) that could be situated in the middle of two NEFs, evoking the tropes of *The great transition* while also stressing the importance of framing.

I started looking at NEF's public interventions between 2007–2013 with a hypothesis: particular intellectual positions and resources shape how an organisation responds to the economic crisis. This conjecture was based on the fact that it is unlikely that individuals—let alone institutions—that have an established position in relation to relevant audiences and supporters will engage in the risky process of repositioning, especially around core concerns to their identity such as their views on the economy. Intellectual change risks alienating allies and think tanks without allies do not survive long. The moments of this crisis, and the expectable content of their work, were illustrated in Chapter 2 in a model that, for NEF's case as a heterodox think tank critical of the UK's economic model, presented this tentative series of public interventions (along with a timeframe and examples):

- 'The end is nigh' (2007–2008): *The Coming First World Debt Crisis* (Pettifor 2006).
- 'We told you so' (September–December 2008): *From the Ashes of the Crash* (Simms 11/2008).
- 'This is an opportunity to turn things around': (2009–2010 general election): *The Great Transition* (NEF 10/2009).
- 'We should have learned' (2010–2011): *Subverting Safer Finance* (Greenham et al. 03/2011).
- 'This will happen again' (2012–2013): *How Did We Get Here?* (Meadway 02/2012).

To be sure, these quadrants can overlap—a more recent blog post could be framed either in the first or last of these: a crisis is still looming, the question being "not if but when" (Meadway 11/2014). However, they proved to be a fair working assumption. Examples of each of these 'moments' were not difficult to find. The model, did not, however, account for ideational change outside discourses 'about' the crisis. As I have shown, NEF underwent important transformations with regards to how they engaged with the public. These, crucially, arose from reflections over their own role in debates about economics and austerity and are visible in the most self-reflective passages of *Framing the economy*. This implies that NEF needs a degree of self-awareness and a working diagnosis of what is likely to reverberate even while trying to convey its message. On intellectual change, one interviewee commented:

> [The 2008 crash] is so different to other crises. Like after the great depression, why has nothing changed? Historically you get crises, whether it's the great depression you get Keynesianism for instance. Whether it's the oil crisis in the 1970s, you get neoliberalism. You know, there's crisis there's change. And that hasn't happened this time, which is very interesting if you dissociate yourself from it. (NEF interview)

To respond to this enigma, this interviewee referenced Milton Friedman's view of policy change. Friedman saw crises as the main junctures where important transformations are possible, opening 'windows of opportunity' in Kingdon's (2003) parlance. Friedman's uncanny mirror image, Antonio Gramsci, was also mentioned, as were references to the 'battle of ideas' of which NEF is a part. In line with such an understanding of change—as being more likely after critical moments—the 2008 crisis is considered a wasted opportunity. Their adversaries, the defenders of the status quo, rather than in disarray, seemed even better organised after the event.

Why was this window of opportunity missed? It would certainly be more than a little unfair to claim NEF failed to capitalise on it. A think tank is only one actor among many, and broader political tendencies raise a tide they do not control. The 2008 crisis and the subsequent election of a Conservative-led coalition committed to a sweeping austerity programme had little to do with NEF and affected its work profoundly. Yet many interviewees still seemed faintly disappointed, believing that more could have been done but was not. Perhaps this disillusionment could be linked to the need for much of their resources and time to cater to the research priorities of funders, which limits how much they can dedicate themselves to other endeavours. Due to this dependency on project-based funding, there was a sense that access to substantial core support could have made their job easier and enhanced their impact. With such resources, presumably more could have been done to disseminate their ideas across publics and coordinate their message.

In the face of this lingering sense of frustration, and in a political and media environment sometimes hostile to their ideas, the most noticeable intellectual changes in NEF occurred at the level of how they believe the public understanding of economics takes shape, rather than concerning the core principles that guide the organisation. This is why towards 2013 they focused on framing strategies and on generating networks of like-minded people to coordinate and disseminate their thinking. An index of that shift is that, while Schumacherian arguments certainly continue to inform NEF's work, it is difficult to imagine they would be presented these days with the same gaiety with which they were in 2008–2009.

These changes beg the question, how does NEF stand as an intellectual unity in time, with so many centrifugal forces, particularly for an institution so dependent on contract-research funding? There is no straightforward answer to this. Even though there are countless instances where interviewees described the type of society they advocate for, some still pondered aloud what the 'new economy' is and wondered whether they had reached sufficient clarity on the matter. One casually mentioned that once, he heard a think-tanker from another organisation say a certain policy idea seemed 'neffy,' yet he did not agree it did. Sometimes concepts of what an organisation is about do not match how they are perceived by outsiders, and discrepancies can arise even between its own members. NEF has reached a certain recognisability (the concept of 'neffy' makes sense, at least among peer policy experts) yet work remains in terms of coordination and dissemination (its contours are fuzzy). That is, in some sense, NEON's mission, to reach a collective, simple, and cogent definition of the 'new economy' that can convince others, change frames, and be widely disseminated.

It could be said that there is greater clarity over what NEF is against—'old' economics. If that is the case, the 'new economy' could be considered a negative notion, useful for bringing together disparate actors. This might explain the difficulty in proposing an alternative to neoliberal policies, such broad negative, vehicular ideas, according to McLennan (2004: 496) tend to stay at a level of interpretive critique that allows for multiplicity and non-commitment. If this is the case—that it is easier to rally 'against' than to rally 'for'—it is by no means a problem only for NEF but extends more broadly to the left.

In view of all of this, could a think tank such as NEF be studied with similar theories and methods as those applied to public intellectuals, as discussed in previous chapters? I contend that yes, in the widest sense. NEF is certainly more institutionally sensitive to funding climates and other institutional pressures than intellectuals *à la* Sartre, but it shares with them many features. First, since as we have seen, the very business that maintains it afloat is producing public interventions, but also because, as an organisation, NEF can and does learn. In this case, the main lesson of 2008 being, if anything, that to champion an idea whose time has come is not enough.

References

Baert, P. (2012). Positioning theory and intellectual interventions. *Journal for the Theory of Social Behaviour, 42*(3), 304–324.

Baert, P., & Booth, J. (2012). Tensions within the public intellectual: Political interventions from Dreyfus to the new social media. *International Journal of Politics, Culture and Society, 25*(4), 111–126.

BBC. (2009). *Brown calls for 'green new deal.'* Accessed 10 October 2014. http://news.bbc.co.uk/1/hi/uk_politics/7927381.stm.

Boyle, D. (2014). *Broke: Who killed the middle classes?*. London: Fourth Estate.

Bronk, R. (2009). *The romantic economist: Imagination in economics.* Cambridge: Cambridge University Press.

Cronin, B. (2010). The diffusion of heterodox economics. *American Journal of Economics and Sociology, 69*(5), 1475–1494.

CUSPE. (2013). *Stewart Wallis—The flawed dominance of economics*. Accessed 4 November 2014. https://www.youtube.com/watch?v=7TbpQlcWaYQ.

Ekins, P. (1986). *The living economy: A new economics in the making*. London: Routledge.

Eyal, G., & Levy, M. (2013). Economic indicators as public interventions. In T. Mata & S. Medema (Eds.), *The economist as public intellectual* (pp. 220–253). London: Duke University Press.

Financial Times. (2009). Top bankers destroy value, study claims. Accessed 22 March 2016. https://www.ft.com/content/7e3edf6e-e827-11de-8a02-00144feab49a.

Friedmann, J. (1992). *Empowerment: The politics of alternative development*. Hoboken: Wiley-Blackwell.

Greenham, T., Jackson, A., Ryan-Collins, J., & Werner, R. (2012). *Where does money come from: A guide to the UK monetary and banking system*. London: NEF.

González Hernando, M. (2018). Two British think tanks after the global financial crisis: Intellectual and institutional transformations. *Policy & Society, 37*(2), 140–154.

Guardian. (2008a). *Andrew Simms: Tackling the 'triple crunch' with a green new deal*. Accessed 28 October 2014. https://www.theguardian.com/commentisfree/2008/sep/19/creditcrunch.marketturmoil.

Guardian. (2008b). *Andrew Simms: The final countdown*. Accessed 28 October 2014. https://www.theguardian.com/environment/2008/aug/01/climatechange.carbonemissions.

Guardian. (2010). *Anna Coote: A shorter working week would benefit society*. Accessed 30 October 2014. http://www.theguardian.com/commentisfree/2010/jul/30/short-working-week-benfit-society.

Guardian. (2014). *Stewart Wallis: An economic system that supports people and planet is still possible*. Accessed 4 November 2014. http://www.theguardian.com/sustainable-business/2014/nov/04/economic-system-supports-people-planet-possible.

Harré, R., Moghaddam, F. M., Pilkerton Cairnie, T., Rothbart, D., & Sabat, S. R. (2009). Recent advances in positioning theory. *Theory and Psychology, 19*(5), 5–31.

Jackson, T. (2011 [2009]). *Prosperity without growth: Economics for a finite planet*. London: Routledge.

Kingdon, J. (2003). *Agendas, alternatives and public policies*. New York: Longman.
Lakoff, G. (2004). *Don't think of an elephant! Know your values and frame the debate*. London: Chelsea Green.
McLennan, G. (2004). Travelling with vehicular ideas: The case of the third way. *Economy & Society, 33*(4), 484–499.
Meadows, D. H., Meadows, D. L., Randers, J., & Behrens, W. W., III. (1972). *The limits to growth: A report for the Club of Rome's project on the predicament of mankind*. New York: Universe Books.
Medvetz, T. (2012). *Think tanks in America*. Chicago: University of Chicago Press.
Mirowski, P. (2013). *Never let a serious crisis go to waste*. London: Verso.
Osborne, T. (2004). On mediators: Intellectuals and the ideas trade in the knowledge society. *Economy & Society, 33*(4), 430–447.
Pettifor, A. (2003). *The real world economic outlook 2003: The legacy of globalization: Debt and deflation*. London: Palgrave.
Pettifor, A. (2006). *The coming first world debt crisis*. London: Palgrave.
Pirie, M. (2012). *Think tank: The history of the Adam Smith Institute*. London: Biteback.
Reuters. (2009). *On wealth versus well-being*. Accessed 20 October 2014. http://blogs.reuters.com/davos/2009/02/01/on-wealth-versus-well-being/.
Schumacher, D. (2011). *Small is beautiful in the XXI century: The legacy of E.F. Schumacher*. Devon: Green Books.
Schumacher, E. F. (1973). *Small is beautiful: Economics as if people mattered*. London: Blond & Briggs.
Simms, A. (2013). *Cancel the apocalypse: A new path to prosperity*. London: Little Brown Book.
Skidelsky, R., Jackson, T., Schor, J., Coote, A., Franklin, J., Harris, M., et al. (2013). *Time on our side: Why we all need a shorter working week*. London: NEF.
Stone, D. (2007). Recycling bins, garbage cans or think tanks? Three myths regarding policy analysis institutes. *Public Administration, 85*(2), 259–278.
TED. (2010). *Nic Marks: The Happy Planet Index*. Accessed 24 October 2014. http://www.ted.com/talks/nic_marks_the_happy_planet_index.
Wolf, M. (2014). *The shifts and the shocks: What we've learned—And have still to learn—From the financial crisis*. London: Penguin.

Think Tank Reports and Blog Posts (NEF, available at neweconomics.org)

Afoko, C., & Vockins, D. (08/2013). *Framing the economy.*

Boyle, D., Johnson, V., Walker, P., & Wimbush, A. (05/2009). *Future news.*

Boyle, D., Nissan, S., & Spratt, S. (03/2009). *I.O.U.K.*

Brown, J., & Nissan, S. (06/2007). *Reconsidering UK community development finance.*

Chowla, P., Johnson, V., & Simms, A. (01/2010). *Growth isn't possible.*

Coote, A. (10/2010). *Cutting it: The big society and the new austerity.*

Coote, A. (04/2012). *The wisdom of prevention.*

Coote, A. (10/2012). *Beyond Beveridge: A new economics vision of a new social settlement.*

Coote, A., & Franklin, J. (02/2009). *Green well fair.*

Coote, A., Franklin, J., & Simms, A. (02/2010). *21 hours.*

Coote, A., Ryan-Collins, J., & Stephens, L. (11/2008). *The new wealth of time.*

Esteban, A. (10/2012). *Natural solutions: Nature's role in delivering well-being and key policy goals—Opportunities for the third sector.*

Green New Deal Group. (07/2008). *A green new deal: Joined-up policies to solve the triple crunch of the credit crisis, climate change and high oil prices.*

Green New Deal Group. (12/2009). *The cuts won't work.*

Green New Deal Group. (09/2013). *A national plan for the UK: From austerity to the age of the green new deal.*

Greenham, T., Potts, R., Prieg, L., & Simms, A. (02/2011). *Featherbedding financial services.*

Greenham, T., Potts, R., Prieg, L., & Simms, A. (03/2011). *Subverting safer finance: How the UK holds back global financial regulation.*

Greenham, T., Prieg, L., & Ryan-Collins, J. (11/2011). *Quid pro quo: Redressing the privileges of the banking industry.*

Huber, J., & Robertson, J. (06/2000). *Creating new money: A monetary reform for the information age.*

Johnson, V., Simms, A., & Smith, J. (10/2007). *Chinadependence.*

Kersley, H., & Steed, S. (12/2009). *A bit rich.*

Kjell, P., Simms, A., & Potts, R. (06/2005). *Clone town Britain.*

LSE. (2012). *About time: Examining the case for a shorter working week.* Accessed 21 April 2015. https://www.youtube.com/watch?v=nqI951u9emQ.

Meadway, J. (11/2011). *Audio: James Meadway addresses occupy London.*

Meadway, J. (02/2012). *How did we get here?*

Meadway, J. (04/2013). *Why we need a new macroeconomic strategy.*
Meadway, J. (11/2014). *The next financial crisis—Not if but when.*
Mitchell, S., Neitzert, E., & Shaheen, F. (10/2010). *Why the cap won't fit.*
Muttitt, G. (02/2003). *Degrees of capture: Universities, the oil industry and climate change.*
NEF. (07/2001). *Timebanks: A radical manifesto for the UK.*
NEF. (05/2009) *Tools for you: Approaches to proving and improving for charities, voluntary organisations and social enterprise.*
NEF. (10/2009). *The great transition.*
NEF. (12/2010). *Response to independent commission on banking issues paper.*
Nissan, S., & Thiel, V. (09/2008). *UK CDFIs—From surviving to thriving.*
Penny, J. (07/2012). *Cutting it in Birmingham: Why the grass roots aren't growing any more.*
Penny, J., & Slay, J. (08/2013). *Surviving austerity.*
Potts, R. (Ed.). (11/2008). *Triple crunch.*
Reid, S. (04/2013). *Mythbusters: Britain is broke—We can't afford to invest.*
Sharpe, R. (06/2010). *An inconvenient sandwich: The throwaway economics of takeaway food.*
Simms, A. (11/2008). *From the ashes of the crash.*
Simms, A. (11/2008). *Nine meals from anarchy.*
Simms, A., & Conisbee, M. (10/2012). *National gardening leave: Why Britain would be better off if we all spent less time at the office.*
Wimbush, A. (02/2010). *The bigger picture.*
Yang, Y. (06/2013). *Rethinking economics: A new conference this weekend.*

4

The Adam Smith Institute: The Free Market's Praetorian Guard

The Adam Smith Institute (ASI) is one of the most recognisable think tanks in Britain. Established in 1977 by three St. Andrews University graduates—Madsen Pirie and brothers Stuart and Eamonn Butler—its name honours the bicentennial of *The Wealth of Nations* (1776), the Scottish heritage of its founders and the classical-liberal outlook they believed Britain desperately needed. Since then, the ASI can arguably be counted among the most famous and influential free market policy institutes in British history, without ever employing more than ten full-time members of staff.

After four decades, the heads of the Institute are still two of its founders, President Madsen Pirie and Director Eamonn Butler.[1] Both Pirie and Butler are part of the Mont-Pèlerin Society (since 1976 and 1984 respectively, and the latter is also its Vice-President at the time of writing), a free market organisation counting among its former members Friedrich Hayek, Karl Popper, George Stigler, James Buchanan,

[1]Stuart Butler left the ASI to join the Heritage Foundation in 1979, the most important think tank linked to the US Republican Party.

M. González Hernando, *British Think Tanks After the 2008 Global Financial Crisis*, Palgrave Studies in Science, Knowledge and Policy,
https://doi.org/10.1007/978-3-030-20370-2_4

and Milton Friedman. The ASI is often considered pivotal in the history of neoliberal thought (Cockett 1995) and Pirie and Butler are its old guard and recognisable face, although the organisation would not be possible without a broad network of fellows, academics, politicians, interns, and businesspeople.

ASI's history is inextricably linked to the Thatcher premiership. Along with the Institute of Economic Affairs (IEA) and the Centre for Policy Studies (CPS), it formed part of a troika of think tanks that lay the intellectual foundations underpinning the 1980s turn towards market deregulation and welfare reform. According to the literature (Desai 1994; Jackson 2012), while the IEA provided academic heft and the CPS political links, the ASI and its fellows devised policy proposals that dared to advance hitherto unthinkable reforms. This depended on two of its features: first, on their self-understanding as 'policy-engineers' (Pirie 1988; Heffernan 1996), designers of simple and workable policy proposals that could be put in practice to advance a free market agenda; second, on their party non-affiliation. According to Denham and Garnett (1998)—and Pirie himself (2012a: 105)—being at arm's length from politicians allowed the ASI to propound radical reforms that would render most parties' manifestos mild in comparison, setting the groundwork to make less ambitious free market policies seem more moderate. An oft-cited remark by Pirie summarises this vividly: "We propose things which people regard as being on the edge of lunacy. The next thing you know, they're on the edge of policy" (2012a: Backcover).

For these reasons, the ASI is one of the most researched think tanks in Britain, particularly in studies dealing with the 1980s neoliberal reforms and the impact of free market thinkers during that period (James 1993; Heffernan 1996; Denham and Garnett 1998; Jackson 2012; Kay et al. 2013; Djelic 2014). Hence, it has become a salient case-study for the relationship between think tanks and 'policy paradigm' change, especially regarding the swing from the 'post-war' to the 'neoliberal' consensus (Desai 1994). Pirie (2012a) himself published a book, simply titled *Think Tank*, that provides a zestful account of ASI's history: from its humble beginnings in St. Andrews' Conservative Society through the Thatcher, Major, and New Labour years until the dawn of the Coalition government. Across these decades, the ASI has

Table 4.1 ASI's position in *Global 'go-to'* think tank rankings (McGann 2009–2015)

Global 'go-to' ranking category	2008	2009	2010	2011	2012	2013	2014
Worldwide (US and non-US)		24	19	20			
Worldwide (non-US)	10	7	8	8	70	69	69
TTs in Western Europe	8	2	6	9	22	21	16
Domestic economic policy			9	7	8	3	3
International economic policy	5	3		21	11	8	5
Policy-oriented research				17	17	13	12
Impact on public policy			20	23	18	20	24
Social policy		6					
Best use of social media (est. 2013)						16/32	17
Public engagement (est. 2013)						49	40

been a powerful advocate for privatisation, flat taxes, deregulation, voucher systems, and low tariffs. To back these ideas, it has relied on Monetarism, Public Choice theory, Classical Liberalism, and Austrian economics. Although not alone in this mission, ASI's vigorous communications strategy earned it a strong public profile, as it has sought to move decision-making from politicians and civil servants to individuals and the market—linking mistrust from state experts and public servants to a *laissez-faire* view of economics and life choices.

The ASI has, arguably, greater clout in the policy debate than its size would suggest. The Institute has featured in top-ten positions in every edition of the University of Pennsylvania's think tank ranking (Table 4.1), surpassing institutions with much larger budgets and output. Notwithstanding the inconsistencies,[2] oversights,[3] and critiques these rankings have garnered (Seiler and Wohlrabe 2010; Trevisan 2012), they serve as testament to ASI's standing.

Organisational and Funding Structure

ASI's headquarters are located in 23 Great Smith Street, indicated by a small brass plate, a few metres away from Westminster Abbey and the government departments of Education and Business, Innovation and Skills (BIS). In contrast to its prime location and the impressive building in which they are housed, the premises themselves are rather modest. There are a bookshelf and a branded banner, a small flight of stairs, and a few desktop computers behind a folding screen, mostly used by interns and gap-year students. Images of the Institute's namesake adorn the walls. Upstairs there is a small mezzanine, used to host events such as 'Power Lunches,' monthly gatherings, talks, and media interviews. When I first went there, the most senior member present was not yet thirty. Nor Pirie nor Butler work at 23 Great Smith Street; they are

[2] In 2010 the ASI was ranked 8th as the best think tank Worldwide (non-US) and 9th in Western Europe.

[3] In 2013 the ASI was ranked both 16th and 32nd in the same category (best use of social media).

based at a nearby office, leaving the headquarters for meetings and public events.

ASI's reliance on interns is reminiscent of its distinctive focus on the young. In the Institute's early days, it depended on the financial contribution of its educational ventures and the teaching salaries of its founders. Nowadays that legacy continues through several instances. One is 'Freedom week' (co-organised with the IEA), a week-long seminar series aimed at university students, with core readings covering *A beginner's guide to liberty* (Wellings 12/2009), Hayek, Smith, and others. One could also mention that the ASI provides support for libertarian student societies and organises school visits, seminar series,[4] essay prizes,[5] and 'The Next Generation' (TNG): a club of under-30 free-marketeers who meet once a month for a drinks reception and a short speech by a prominent ally (more on this later). Over the years, outreach projects such as these have allowed the ASI to build a network of like-minded people, many of whom have at some point contributed as interns, researchers, donors, fellows, or writers, either for the ASI or for other free-market organisations (Pirie 2012a: 20).

However, the ASI's *raison d'être* is, of course, changing public policy. To achieve this, the Institute seems to embody the proverbial 'Medvetzsian' think tank: it invokes resources from Medvetz's four fields (2012a) in order to influence the political debate—media, politics, academia, and the economy. Indeed, Pirie's (op. cit.) book is awash with references to their networks, which span academics, public intellectuals (Friedrich von Hayek himself was an ASI trustee), journalists, politicians, businesspeople, philanthropists, and think-tankers in the UK and abroad. In Pirie's narrative, it is to a large degree these connections what makes the ASI possible.

ASI's heavy reliance on its informal or quasi-formal links is worth emphasising. Unlike many other think tanks, it does not pursue

[4]'Independent Seminar on the Open Society' and 'Liberty Lectures,' for sixth-form (High School) and university students respectively.

[5]The 'Young Writer on Liberty' is a writing competition for under-21, which asks contestants to submit entries on topics such as "Three policy choices to make the UK freer, richer and happier," accessed 15 March 2015, http://www.adamsmith.org/student-outreach/young-writer-on-liberty/.

commissioned-research grants (Clark 2012), nor is it subdivided into teams. Instead, it depends on a number of 'fellows,' 'senior fellows,'[6] and other sympathisers to publish under its label, either in the form of blog posts, policy briefs, or lengthier reports. Many of these collaborators are former ASI employees—Tim Evans, Tom Clougherty, JP Flouru. Some are full-time academics—Richard Teather,[7] James Stanfield,[8] and Tim Ambler[9]—while others are market analysts and advisors, bankers, and lawyers, either supplying written content or providing advice—Gabriel Stein, Anthony Evans, Preston Byrne, and Nigel Hawkins. ASI fellows also include journalists—James Bartholomew—and members of the business and finance world—Lars Christensen, Miles Saltiel, and Tim Worstall. Many are also linked to other think tanks and advocacy organisations, including Christopher Snowdon,[10] Deepak Lal,[11] Keith Boyfield, and Jamie Whyte (IEA members, writers, and fellows), Dominique Lazanski and Eben Wilson (TaxPayers' Alliance), Anton Howes (Liberty League), and Tom Papworth (CentreForum).

These connections are often reciprocated. The IEA regularly publishes primers by Eammon Butler, and both him and Madsen Pirie were counted (while the list was public) as members of the TaxPayers' Alliance Academic Advisory Council.[12] The Institute is also part of transnational free market groupings, such as the Atlas Network (Djelic 2014) and, until 2009, the Stockholm Network, while also sharing

[6]Similarly to NEF, 'fellows' are people linked to the ASI who share its general philosophy and contribute to the organisation without being involved in its everyday operations. As one interviewee puts it, "'fellows' are those who benefit from being associated to the ASI, and 'Senior Fellows' are those the ASI benefits from being associated with" (ASI interview).

[7]Richard Theater is Senior Lecturer in Tax Law at Bournemouth University.

[8]James Stanfield is Director of Development at E.G. West Centre in Newcastle University.

[9]Tim Ambler is a retired Senior Research Fellow in Marketing at the London Business School.

[10]Christopher Snowdon is Director of Lifestyle Economics at the IEA.

[11]Deepak Lal is Emeritus Professor at UCLA and Fellow of the CATO Institute. He is also a frequent contributor to the IEA.

[12]Accessed 25 March 2015, https://web.archive.org/web/20070621095655/http://www.taxpayersalliance.com/about/advisory_council.php.

links with several libertarian think tanks from the US.[13] Indeed, these contacts allow the ASI to access a complex of sympathetic experts and institutions they can cite, invite as speakers or be cited and invited by, rendering the intellectual networks and their public interventions appear more robust than they otherwise would.

In relation to the Institute's day-to-day operations, most ASI fellows and senior fellows do not collaborate on a regular basis and receive little or no financial reward when they do. As such, and given the paucity of exclusively research-oriented staff and funding, the ASI tends to produce fewer policy reports than larger generalist think tanks. It concentrates instead on an active online presence and an aggressive communications strategy, aimed at increasing the impact of their publications and events. ASI staff and fellows routinely appear in television, radio, broadsheets, and tabloids, including the *Financial Times*, *The Sun*, *The Daily Mail*, *The Spectator*, *The Times*, *The Daily Telegraph*, and *The Guardian*. It is also worth noting that several current and former ASI members have worked as columnist and editors for *CityAM* (a free, business-focused newspaper distributed across the City of London).

Politically, although the ASI declares itself to be party-independent, it has a long history of collaborating with the Conservatives, and to a lesser extent with market-friendly member of New Labour and the Liberal-Democrats. However, it is not considered to be the closest think tank to the Conservative leadership (Pautz 2012b), though it does maintain contact with senior Conservatives (e.g., Michael Forsyth, Michael Portillo, John Redwood) and a younger generation of libertarian-minded Tories. To build these networks, the ASI relies on years of building and maintaining personal relationships and hosting informal events, such as 'power lunches'—i.e., meals on ASI's premises where politicians, members of the media, and free-market sympathisers are invited to hear a short speech by

[13]One could mention ASI's links with Edwin Feulner and Stuart Butler, respectively former President and Director of the Center for Policy Innovation of the Heritage Foundation. To these one could add the frequent visits of guest speakers from CATO Institute and George Mason University, Madsen and Eamonn's past bonds to Hillsdale College, and the public lectures held by ASI members for US-based libertarian think tanks such as the now defunct National Center for Policy Analysis (NCPA).

a kindred thinker. Instances such as these speak of a well-honed impact strategy, devised over decades of proximity to Westminster. One interviewee's remarks are worth quoting in length:

> So we have [experts we agree with] over, and when they speak we invite the financial journalists, […] the business editors of newspapers, we sit them around a table […] The aim is to influence them, because what they write then influences the Treasury and the ministerial team, and they begin to feel they're isolated unless they go along with what the consensus view emerging from the top experts says. So we seek to influence political events through the public domain; we don't do it in private, we do it publicly […] There's no point in trying to do it privately because if you can change a minister's mind in a half-hour meeting on Tuesday someone else can change it back on Wednesday, but if you've influenced the tide of opinion, that's much more difficult to change back. (ASI interview)

The sum of these endeavours speaks of a small yet well-connected think tank that seeks to secure as large an impact with as few resources as possible. Indeed, although there have been noticeable organisational changes in the years following the crisis, the Institute has remained relatively low-budget. Relying as it does on a network of allies, ASI's public presence is much more imposing than its office or staff numbers would suggest. As an annual review puts it, "[m]any think tanks have much bigger budgets, but few deliver as much 'bang for the buck' as the Adam Smith Institute" (ASI 01/2011).

But where does the 'buck' come from? Unlike most British think tanks, the ASI is not formally a charity. It ceased being so in 1991.[14] Nowadays, it is composed of a few formal organisations, some of which are registered in the Charity Commission. On the reasons why, one interviewee commented:

> Our activities do qualify legitimately as charitable, but […] the public at large does not understand that, and we didn't want to have people saying

[14]Data from Charity Commission (Reg. No. 282164), accessed 18 February 2015, http://apps.charitycommission.gov.uk/Showcharity/RegisterOfCharities/RemovedCharityMain.aspx?RegisteredCharityNumber=282164&SubsidiaryNumber=0.

> 'look, they're pretending to be a Charity but in fact they're advocating free market libertarian ideas.' Well that's what we are! [...] The public thinks we are hypocritical and deceitful pretending to be a charity, so we don't pretend to be a charity. (ASI interview)

Currently, the Institute is mainly divided in the ASI (Research) Ltd. and the Adam Smith Research Trust. According to email communications with an ASI member (09 March 2015), the first is a non-profit company limited by shares that most of ASI's work operates under. The second is a registered charity that funds ASI (Research) Ltd.'s non-political educational work.[15] Pirie (2012a: 34–35) himself gives a description of this peculiar institutional architecture in his book, which took years to set up. To these bodies, one should add a few smaller charities, many of which are discontinued, used to assign funds to educational ventures.

It is, however, difficult to speak much about ASI's sources of income, as the Institute has a strict policy of non-disclosure. The available data on ASI (Research) Ltd. and Adam Smith Research Trust is summarised (Table 4.2). Their numbers do not reveal much—as the deficits ASI (Research) Ltd. operates under are not a good indication of the Institute's budget—but I report them nonetheless for two reasons. First, because these are the publicly available figures of ASI's finances. Second, to confirm the comparatively small size of the operation when put beside most generalist think tanks in the UK, whose numbers ordinarily run in the millions.

In personal correspondence (09 March 2015), one ASI member stated that, in any given year, about 50% of its funding comes from charitable foundations, 35% from individuals, and 15% from corporate donors. ASI's website details the modalities of some of those donations—separated in branches from 'members' to 'benefactors.'

[15]See the charity Foundation for Research and Education in Economics (Reg. No. 270958), headed by Eamonn Butler. Its mission, as reported in its annual accounts, is "the funding of research, conferences and publications in the general field of social sciences." According to personal communications with an ASI member (9 March 2015), this entails fundraising for smaller bodies such as student organisations, accessed 10 March 2015, http://apps.charitycommission.gov.uk/Showcharity/RegisterOfCharities/CharityWithoutPartB.aspx?RegisteredCharityNumber=270958&SubsidiaryNumber=0.

Table 4.2 A.S.I. (Research) Ltd. and Adam Smith Research Trust financial overview

ASI (Research) Ltd.	2007	2008	2009	2010	2011	2012	2013
Total assets	46,652	42,815	39,656	25,686	34,178	40,884	28,319
Debts	235,033	204,282	109,986	75,054	56,211	56,855	80,495
Balance	−188,381	−161,467	−70,330	−49,387	−22,033	−15,971	−52,176
Adam Smith Research Trust	*2007*	*2008*	*2009*	*2010*	*2011*	*2012*	*2013*
Income	11,440	34,583	38,735	29,288	206,209	45,203	212,577
Expenses	31,916	89,643	62,511	55,226	39,075	47,605	67,630
Net	−28,561	−55,209	−23,856	−26,025	−12,908	−2451	144,870
Assets	182,356	153,795	98,586	74,730	49,453	36,545	34,904
Balance	153,795	98,586	74,730	49,453	36,545	34,094	178,964

Data retrieved from the Companies House (Reg. No. 1553005 and 802750), accessed 28 February 2015, http://wck2.companieshouse.gov.uk//compdetails and http://apps.charitycommission.gov.uk/Showcharity/RegisterOfCharities/CharityWithoutPartB.aspx?RegisteredCharityNumber=802750&SubsidiaryNumber=0

The ASI is also registered as a non-profit foundation in the US, and contributions to the Institute are tax-deductible under the 501(c)(3) exemption. In terms of total budget, ASI's resources grew between 2007 and 2013, although precise data to illustrate this is difficult to find. One interviewee commented:

> Our target next year [2015], I don't think this is any secret, is to raise our budget to £330,000, meaning it's below that now […] The largest amount we spend is on staff salaries. It doesn't include [Madsen nor Eammon]. [Other think tanks] can afford to do much bigger projects than we can. We have to be incredibly lean and cost-effective. (ASI interview)

The mystery surrounding ASI's sources has given ammunition to its critics.[16] The Institute has not, however, avoided the issue. In 2012 an ASI fellow, in response to ASI's poor rating in a think tank transparency survey,[17] wrote: "[w]e are delighted to be judged on the quality of our ideas and our arguments. And as a non-party political, independent and non-profit think tank that's all that we should be judged on" (*Daily Telegraph* 2012). On the same issue, an interviewee said:

> We don't publish [donors'] names, and that is for two reasons. One is to protect their privacy. The other is actually more serious. We don't want them intimidated. If you're known to be a donor of an organisation like the ASI you risk having groups like Occupy coming and staging demonstrations […] You do risk intimidation by those groups who oppose what we stand for. (ASI interview)

[16] This opacity possibly explains a common confusion. The public accounts of Adam Smith International—an international development organisation to which ASI members were formerly affiliated but which Pirie and Butler claim is now completely independent—are often conflated with those of the ASI. The Adam Smith International budget goes well into the millions of pounds, a significant part of it coming from the DfID (*Guardian* 2012). See also accessed 13 March 2015, http://www.adamsmithinternational.com/about-us/our-history/.

[17] Accessed 20 March 2015, http://whofundsyou.org/org/adam-smith-institute.

Even setting aside these blind spots, ASI's low expenses and substantial proportion of core funding[18] suggest it could compete with larger contract-dependent think tanks by being more responsive, if having less capacity to pursue long-term research. And although critics speculate to what extent do the priorities of donors shape the Institute's public interventions or the areas they choose to work on (Kay et al. 2013), the very fact the ASI advocates for deregulation and privatisation is likely to entice corporate support. In 2013, it was reported by *The Observer* (2013) that the ASI received £13,000 from the tobacco industry. One interviewee commented:

> A constant moan of the left is that we're funded by Big Tobacco […] in order to, you know, defend smoking […]. The left gets it completely the wrong way round. It's because we advocate individual choice that we come out against plain packaging […]. It's because we're so robustly libertarian that these organisations come along and give us money. (ASI interview)

Following the above, it is conceivable that ASI's choice of what to intervene in has been more flexible than that of think tanks heavily dependent on research contracts, among other reasons because the ASI conducts little to no in-house research in the first place. Further, to maintain independence, ASI reports to have a cap on the percentage of its total budget that can come from any given donor (Heffernan 1996; Clark 2012). Hence, presumably the Institute had between 2007 and 2013, compared to commission-oriented think tanks like NEF, more space to set their own priorities, if having to rely to a greater extent on external authors.

[18]"Nearly all [donations are] core funding. Some people say 'I'd like to spend on your youth activities,' but almost never for a specific event […] We do take individual donations for some of our projects [for instance] 'we are hoping, with the IEA to run a Freedom Week again, could you possibly support us?' […] but normally it's either core funding or for our youth activities […] They might specify" (ASI interview).

Style and Tropes

Before exploring ASI's most common tropes, a few caveats should be mentioned. It was argued previously that, in order to describe a generalist think tank's purported 'brand,' it is more fruitful to focus on 'tropes' and recurrent arguments than on the research areas it focuses on, as these may vary with the political climate and available funding. This certainly continues to apply. However, given ASI's substantial proportion of core funding, their choice of policy area can be more flexible than for larger think tanks and those more reliant on research contracts, if having fewer resources to pursue longer-term empirical research.

Additionally, one should remember that free-market economics and neoliberalism are burdensome concepts. For that reason, it is problematic to argue that the Institute represents or reflects all of the 'neoliberal thought collective' (Dean 2012), even if the ASI has contributed to the definition of what neoliberalism means in public policy and is closely associated to the two most important international coalitions of that persuasion, the Mont-Pèlerin Society and the Atlas Network. Thus, although there has been much work on defining the contours and history of neoliberal thought, in order to assess ASI's specific contribution, one must to an extent start anew.

Furthermore, because of ASI's organisational model—which relies heavily on external experts and fellows—the format of public intervention that should be considered representative is different from that of research contract-oriented think tanks. This means that in ASI's case, policy papers are arguably not the central node that epitomises and brings together the Institute's thinking. One interviewee comments:

> One of the interesting things about the ASI [...] is that most of our policy research is written by outside authors, so we commission them, we edit them and everything, but, to a certain extent, we are constrained by the authors we find and their approach to the issue, whereas if you look at the blog where we're writing ourselves [...] you might get a true reflection of what we were thinking internally. (ASI interview)

ASI's blog, among the most active of any British think tank at the time, is updated at least daily and is the mainstay of its website. It relies on regular contributions by Butler, Pirie, senior researchers, interns, and fellows, as well as re-publishing outside pieces by ASI authors. Although indicative of the think tank's intellectual changes, its focus tends to be driven by daily events, just as many of their interventions in the media. An interviewee states, the topics covered by the blog are:

> largely opportunistic, based on what people are talking about. In a way it does promote our long-term goals occasionally, but its main purpose is to give a day-to-day response on what it's happening [...]. It has to be current [...]. So when someone like Piketty brings out a book and everyone is reading about it and making it a *New York Times* best-seller, we have to talk about it too. (ASI interview)

With those caveats in mind, one could say ASI's modality of public intervention has often been that of the 'defender of the order' (see Sapiro 2009). Yet, that order is not necessarily the status quo *tout court*, but one derived from a specifically Anglo-American intellectual heritage: Classical Liberalism, based on individual freedom and suspicion of state action. It is an 'order' based on a vision of how things *really are*, or at least how they would be without government intervention, and which confronts its custodians against other figures of authority. Hence ASI's need to garner a type of credibility that supersedes that of elected or appointed officials, as well as that of more specialised experts.

A somewhat old-fashioned front aids their pursuit of such authority, typified by the use of anchors and symbolically charged props. Routinely, members of the ASI—traditionally meaning Pirie and Butler—have sought, against what they see as a more emotional left, to portray themselves as piercing and commonsensical. A former secretary of Mensa (a famous high IQ society), Pirie has published books on logical fallacies and how to win arguments (Pirie 2007), released a YouTube video series on the same topic, and has as a personal trademark to sport a bowtie (Pirie 2012a: 53).

As with other think tanks, ideas and institutional forms tend to coalesce over time. Instances of this include how ASI's longstanding admiration for the US and its associated entrepreneurial spirit informs frequent cooperation with institutions and individuals from that country (Pirie 2012a: 4, 55). Further, one of their oldest initiatives has been to publish a yearly calculation of the given day of the year in which the average wage-earner finishes paying her fiscal contribution, the *Tax Freedom Day*. In another example, even if the Institute's claim to represent the thought of Adam Smith has been contested by its ideological adversaries (e.g., Tax Justice Network 2010), Madsen Pirie and others were key for the installation of a statue of the Scottish economist in Edinburgh city centre.

This form of public engagement has developed into a recognisable and frequently confrontational style. This is noticeable in their public interventions in most policy areas, from macroeconomics to the taxation of unhealthy food (BBC 2013), even when debating more specialised opposition (Worstall 07/2010). ASI has directed rejoinders to targets as varied as the British Medical Association (Hill 02/2012), the Archbishop of Canterbury (*Daily Telegraph* 2009b), vegetarians (*Daily Telegraph* 2009c) and environmentalists (Worstall 2012). Tim Worstall, one of ASI's most active fellows, has written scathing posts against several figures broadly associated with the left and academia, from Richard Murphy of the Tax Justice Network to columnist Owen Jones, Professor Lawrence King from the University of Cambridge, Lord Skidelsky, Adair Turner, and NEF. This type of performance has many antecedents in the history of the organisation. Pirie states in his book:

> We were forceful advocates and spoke like true believers, because this is what we were [...] we wanted to convince public opinion [...] about the superiority of markets, choices and incentives, and the merits of privatization. This required us to adopt a high profile and a somewhat combative one. (Pirie 2012a: 86–87)

ASI's denunciations of rival intellectuals often claim that their adversaries do not know what a free market is or how it operates. The rhetorical fulcrum of this argument frequently relies upon the Hayekian assertion that nobody can know the economy and society better than the market—the sum of actions by individuals with local knowledge seeking their self-interest coordinated by the price mechanism. Quantification and grand programmes by privileged actors are, in this view, most often counterproductive. It is, according to the ASI, best to leave decision-making to individuals than to authorities, and the economy to the market's spontaneous order than to state planning. Perhaps foreshadowing what later came to be called 'post-truth politics,' these ideas allow the ASI to argue that others are wrong—even those from areas in which they do not possess in-house expertise—by casting doubts over the very foundations of their judgement and jurisdiction (see Davies and McGoey 2012; McGoey 2012; Mirowski 2013). In such a way, the ASI can intervene in almost any policy area by undermining the knowledge-claims of their political and intellectual adversaries (Jacques et al. 2008). In 2013, ASI and IEA fellow Jamie Whyte published *Quack Policy* (2013), where he rails against evidence-based policymaking based on similar arguments.

This form of argumentation is predicated on a specific view of economics. It is one that tends to invoke a disciplinary 'core' beyond reasonable dispute. This kernel is epitomised by some of the writings of Adam Smith himself, and those failing to heed its insights are sometimes dismissed as 'not true' economists (Worstall 09/2012). The dismal science is, for ASI authors more closely associated with the Austrian school, structured by a view of the nature of exchange and the unplanned order that stems from it. For these reasons, ASI members are often sceptical of complex models and of our ability to plan, wishing to contract rather than expand—as NEF researchers would—the scope of what is considered sound economics (Butler 06/2011).

This understanding of economics is linked to a mode of engagement with its publics that strives to be uncomplicated and accessible (Pirie 2009, 2012b). In line with ASI's educational mission, the Institute has published several primers on canonical free-market thinkers, a majority written by Butler (2007, 2010b, c, 2011a, 2012a). Pirie (2008) has also contributed with *Freedom 101*, where he takes issue with 101 arguments against market liberalisation. It is noteworthy, however, that although pursuing clarity, the line of reasoning in these texts seems at times counterintuitive, a recurrent trope being that well-intentioned actors tend to produce worse societal outcomes than those narrowly following their self-interest. This last point is apparent in their opposition to *Fairtrade* (Sidwell 02/2008), the curtailment of tax havens (04/2009), and publicly funded higher education (Stanfield 03/2010).

Another typical aspect of much of ASI's work is its anti-political inclination (Butler and Teather 04/2009; Pirie 2012a: 4). Derived from a mistrust of experts and politicians, and from a wariness regarding what they see as economic populism (Butler 05/2012), there is an implicit challenge to democracy in their publications. This is informed by the notions that politicians and public servants mostly pursue their own selfish aims (drawing from Public Choice theory) and that individuals are inevitably ignorant on matters of macroeconomics and public policy (frequent in Hayek's sympathisers). What is more, ASI members have made reference to a specific theory of political cycles: pushes for a larger welfare state often end up producing clientelism, brought about by complacency, and sluggish growth ensues. Then radical measures become necessary to reduce public spending and, after a while, the economy recovers and the cycle begins anew. This view even colours ASI's view of the 2008 crisis. One interviewee told me:

> [Between India and China] In the long term I think China is probably the better bet economically. And the reason is that India is a democracy, and in a democracy, periodically, when you do well economically people get complacent, and they turn to a left-wing party which promises more redistribution, and because there are more poor people [...] they'll have a majority for that party, hand out goodies to the people they promised them to, ruin the economy, and about eight years later re-elect the guys who make it sensible again. And in some sense in Britain, Tories rebuild the economy, and then people get complacent and they vote Labour in. Gordon Brown spends everything and produces massive deficits and debts, and then we put the Tories again to fix it. You know there's this [...] cyclical thing in a democracy. (ASI interview)

Keynesians at the Gates: Defending 'Liberty' Against Its Foes

In the setting up of this book, I expected that think tanks' public interventions were likely to be influenced by their previous work. This conjecture implied, in a nutshell, that a financial crisis would put a policy institute that vigorously supports free markets on the defensive, which would elicit a blame-seeking riposte, exonerating the market itself by separating the 'event' from the 'structure.' One finds similar arguments in the scholarship on the matter (Jacobs and Townsley 2011: 188–189; Schmidt and Thatcher 2013). These ideas could also be complemented by a view of the dynamics of intellectual crises set up by Baert and Morgan (2015). They argue that in instances of intellectual discord there is often also a conflict over the perceived seriousness of the matter—with emerging intellectuals stressing it while establish ones tending to downplay it. Debates can be construed as trivial or existential; either as part of the normal functioning of an order or invoking the realm of the 'sacred.' In this interpretation, the 2008 crisis could be seen by defenders of *laissez-faire* liberalism as an illustration of the self-correcting mechanism of free markets simply doing its job. In November 2008, Butler said to *The Guardian*:

> The crisis we've been hit by has been caused by extreme public policy in the US, but it won't necessarily last for years and years. Right now, there's naturally plenty of worry about markets being volatile but by 2011, we'll be back to where we are today. (Butler, in *Guardian* 2008a)

A belittling of the magnitude of the crisis was not the most common response, however. As the severity of the situation became apparent, there was little time for abeyance. Moreover, since the type of order the ASI defends is not necessarily an actually existing one but an ideal, contesting *what* was in crisis was of prime political importance. The Institute's blog was especially active in the weeks following the fall of Lehmann Brothers. ASI members blamed governments and regulators (Butler 10/2008), particularly in relation to the banning of redlining—denying access to credit to those least likely to repay—in the US's housing sector, which they claimed distorted the market and magnified risks (Clougherty 09/2008).

This line of argumentation was, to be sure, present before September 2008. Earlier that year, Hansen (02/2008) cited Walter Williams from George Mason University—which hosts the Koch-funded Mercatus Center and several free-market economists—warning against the consequences of government intervention in the mortgage market, and in

2007, after the Northern Rock bank run, Butler accused politicians of the debacle (BBC 2007). Being that the possibility to fail is considered a ruling principle of free markets, the situation would only worsen after bailouts and the de facto nationalisation of banks (Butler 09/2008). For the ASI, the crisis was not caused by unfettered greed, but by a series of ill-though incentives and policies (*Guardian* 2008b).

In a rapidly shifting context, the Institute saw a renewed sense of purpose: to defend the then besieged free-market ideas and to step back on the offensive against those challenging them. Statists were at the gates of the City, and "[…] opposing the Keynesian-revival is probably the key economic task for defenders of freedom in 2009" (Clougherty 01/2009). Hence, the financial crisis generated newfound energy and threats, offering opportunities to rally support. ASI's 2009 yearly review, a propitious instance to address potential donors, states:

> Ideas have consequences: and every day, interest groups are bombarding governments with demands for more spending and more regulation. We need think tanks like the Adam Smith Institute to stand up against such demands. (ASI 01/2010: 1)

Organisationally, towards the end of 2008 some important changes were underway. Tim Evans and Tom Clougherty—then, respectively, Consulting Director and Executive Director—sought to professionalise ASI's functioning. This meant having a more permanent group of middle-tier employees between Butler, Pirie, and junior staff, as well as diversifying the Institute's output. To achieve that, they sought to maintain and expand their network of donors and develop a more thorough business plan. In terms of the fundraising effort, one interviewee commented:

> We had […] around 2007 a very loyal donor base that probably had been giving more or less the same amount of money for a long time […] I think it paid dividends […] trying to build relationships with new donors, trying to work out what their interests were within the field of free-market libertarianism and let that to a certain extent guide the areas in which we focused on. But there was really no sense that chasing after money was guiding what we were doing. In fact, it turned out to be the other way round. We wanted to do more on financial issues […] and that

> was where we had great donor opportunities. Our budget [...] doubled in the space of a few years, from a very low base to a relatively low one [...] and allowed us to increase staff [...]. In terms of the 'core' ASI people, it went from about 5 to [...] 9-10 people. (ASI interview)

Publication figures may be one of the indexes of this push for renewal. Butler, one of the Institute's most prolific authors, produced a considerable part of his output following the crisis. At the time of writing, he has published over 28 books in total, with a long hiatus between 1999 and 2005. However, from 2008 until 2013, Butler would pen primers on Mises (2010b), Friedman (2011a), Smith (2011b), Hayek (2012a), and Public Choice theory (2012b), blaming New Labour for the crisis (2009a, b), and advocating for the primacy of free markets more generally (2008, 2010a, 2013).

Outside the organisation, briefly at least, a Keynesian wave seemed to gather momentum (Skousen 02/2009), and free-market advocates were sometimes blamed for their unwillingness to readdress their position in the face of contravening evidence. This ostensible weak flank was defended fiercely, as the ASI sought to shift blame away from capitalism itself—it was, instead, 'a Labour-made' crisis (Butler 03/2009a) and Libertarians had to "fight back" (Butler 03/2009b). This elicited a shift of priorities and at times self-criticism, as there were oversights in the build-up period:

> When the financial crisis was bubbling up we really weren't thinking about those kinds of economic issues at all. I mean hardly anybody was [...] on the free-market side or on the left frankly [...] So I thought we took the eye off the ball a little bit, and if people on the free-market side had been paying more attention to the financial system [...] monetary policy [and] the deficits, which were starting to get bigger at the time, then we would have been more ready [...] more able to give a response faster than it turned out to be the case. But I think everybody made that mistake. (ASI interview)

The circumstances required for the ASI to repeat its message across fora, concentrating a sizeable part of its energy on financial issues (Pirie 2012a: 105). One interviewee told me:

> From doing nothing at all on financial economics on 2006-2007 [...] by 2011-2012 I was doing little else. That was what most of our events revolved around, what we were publishing a lot on, what we got called a lot about by the media, what our donors were most interested in. (ASI interview)

This focus on finance was accompanied by a broader intellectual movement outside the organisation, especially coming from US-based conservative thinkers. The depth of the crisis was reassessed, and its causes traced to errors in fiscal and monetary policy. Against the narrative that irresponsible bankers were the culprits behind the crash, ASI members argued that Fannie Mae and Freddie Mac, public institutions, plus low-interest rates by central banks, were behind the collapse. These ideas were, to be sure, not exclusive to the ASI. It could be argued that, as a think tank depending to such an extent on external authors, their arguments were often similar to those coming from a broader free-market orbit.

Leading up to the 2010 election, the Institute published a series of policy reports, mostly written by fellows, seeking to explain the causes of the financial meltdown and proposing reforms to avoid its repetition. In some of them, the ASI advocated for discrete measures, implying that the causes of the crash could be located in a series of circumscribed factors: loose monetary policy, a distorted housing market, a failed Basel treaty, the oligopoly of rating agencies, and the pernicious incentives faced by 'too big to fail' banks (Ambler 11/2008; Saltiel 03/2009; Boyfield 10/2009). Nonetheless, other reports made a more ambitious argument, questioning the very architecture of the British monetary system (Simpson 06/2009; Redwood 10/2009). Particularly after the publication of Schlichter's (2011) *Paper Money Collapse* and its associated presentation at the ASI[19]—where he explained the recession through the Austrian business cycle theory, blaming elastic money creation by central banks—members of the Institute began pushing against the UK's basic monetary structure, zeroing in on the remit of central banks.

---

[19] Accessed 22 March 2015, https://www.youtube.com/watch?v=jt-6lHTpefU.

In 2010, with the election of a Coalition government committed to fiscal consolidation, the political arena seemed more propitious for the ASI. Public spending cuts, which had been one of the Institute's favoured policy solutions for some time, opened yet another policy area where to push for shrinking the role of the state (e.g., Pirie 08/2009). This political juncture allowed the Institute to recover past work (e.g., Ambler and Boyfield 11/2007), which could be advanced now that Conservatives were in power in the model of the 'garbage can' (Stone 2007). And although it would be simplistic to say that the ASI was influential simply because their priorities and those of the Conservative-led government coincided, this certainly created a favourable atmosphere. Pirie states in his book:

> 2010 marked a turning point in British policymaking, with the election of a coalition government committed to public service reform, civil liberties, and fiscal responsibility. In many areas, the government made a good start, with spending restraint, schools liberalization, and welfare reform dominating the political agenda. But there remains much to be done. (Pirie 2012a: 207)

After the 2010 election, ASI authors frequently used metaphors of excess to castigate what they saw as bloated and unsustainable public spending, which in their view helped bring about the conditions that caused the 2008 debacle. Shortly after the 2010 election, Butler said to *The Guardian* (2010b) that the government needed to go on a 'diet.' These tropes were conspicuous even in the titles of their papers, for instance, in *The party is over* (Hawkins 06/2010) and *On borrowed time* (Saltiel 12/2010). The first proposed average 3% cuts across government departments, while the second argued for a reconsideration of the state provision of healthcare, pensions, and education. Other reports, such as *Rebooting government* (Butler 06/2010) and *Zero base policy* (Pirie 2009), would summarise the policy approach the Institute would advocate for: revisiting the very bases of public policy provision and financing, as if formatting a slow computer. The implications were clear: financial regulation and monetary policy were behind the crisis, an irresponsible expansion of public spending made matters worse, and severe deficit reduction was needed.

In sum, under the Coalition, ASI's public interventions sought to counter any effort to undermine or mitigate the austerity programme. In an annual report, the ASI evoked Thatcher by stating that "[i]n 2011 and beyond, the Adam Smith Institute will be working to ensure that there is no turning back" (ASI 01/2011: 22). Concomitantly, the Institute kept arguing against the idea that financial deregulation was to blame for the 2008 slump. For such reasons, they critiqued the official Turner (Boyfield 03/2009) and ICB reports (Ambler and Saltiel 09/2011) on banking reform. This topic, however, opened a flank for ASI fellow travellers who wished to lambast the coalition's policies from the right, which highlights an internal rift in some of the Institute's work:

> There was concurrently a radical-libertarian response brewing in the minds of people at the ASI which would've entailed much more far-reaching reforms to the way monetary policy and banking were operating, and at the same time we were producing a lot of research which was much more mainstream free market though, much more in step with whatever developments were happening in the political debate. (ASI interview)

In a nutshell, radical responses to the crisis started to originate from within ASI's camp. Some public interventions, for instance, questioned whether Britain is truly a capitalist economy (Whig 11/2011), given the size of government and the existence of a central bank. In that vein, in 2011, the Institute hosted George Selgin, an academic and CATO Institute member who has written extensively for free banking—an economic system bereft of central banking (Selgin 1994). Fittingly, Selgin confronted NEF fellow traveller Lord Skidelsky at an LSE public event (LSE 2011) entitled 'Keynes vs. Hayek,' which epitomised the intellectual debate at the time. The ASI also continued to invite academics of a broadly 'libertarian' persuasion, such as Brendan Brown[20]

[20]Brendan Brown is associated scholar at the Mises Institute.

and Kevin Dowd,[21] albeit "academically qualified libertarians who could write the kind of material we were looking for were in relatively short supply" (ASI interview).

In parallel, in this period, much like in the 1980s, the ASI sought to move the 'Overton window' by arguing the government should go even further in its austerity plans. This is visible in their publication of its 'Budget wishlists.' The areas where ASI publications claimed the market could be liberalised and the state de-scaled included higher education (Stanfield 03/2010), the BBC (Graham 08/2010), the arts (Rawcliffe 03/2010), personal tax (Saltiel and Young 10/2011), and the Post Office (Hawkins 10/2010), many of which drew from years of publications arguing for cuts, deregulation, and voucher systems (e.g., Senior 11/2002).

However, ASI's precise policy influence is difficult to ascertain. The Institute broadly approved of the Coalition's public spending reductions and free market reforms, but their manner and extent were sometimes criticised as too timid. One example of this is their (ultimately unsuccessful) push for profit-making companies to be allowed to run free schools (Croft 04/2011). Besides, past decades had seen the emergence of better-provisioned think tanks on the right, and institutions like Policy Exchange have much closer links to a Conservative leadership that, for a while at least, sought to distance itself from diehard Thatcherism (see Chapter 6). Even so, one instance of policy impact mentioned by ASI interviewees was dissuading the Coalition from raising capital gains tax, through Laffer curve arguments—i.e., higher tax rates result in diminished tax revenue (ASI 05/2010). In this endeavour, they were supported by John Redwood MP and the *Daily Telegraph* (2010). Overall, the Institute continued to be politically active: it was part of a government consultation on welfare reform (DWP 2010), held several Power lunches and organised fringe events at Conservative Party conferences, with presentations from MPs such as John Hutton,

[21]Kevin Dowd is a partner at the monetary policy consultancy Cobden Partners and professor of finance and economics at Durham University.

Douglas Carswell, and, again, John Redwood. Nonetheless, one interviewee claims the ASI enjoyed "a good relationship but not a close relationship" with Conservative frontbenchers.

Concerning institutional changes, as part of the measures brought about by Clougherty and Evans' restructuring, the ASI contracted a handful of more permanent young employees. They brought a new set of preoccupations and skills, and noticeably expanded the breadth of the Institute's output. Among them, we find Charlotte Bowyer,[22] Ben Southwood,[23] Phillip Salter,[24] Kate Andrews,[25] and Sam Bowman. The following paragraphs will focus on the latter.

Bowman is former Deputy Director and is perhaps a natural successor to Pirie and Butler. He is a libertarian of a specific persuasion the ASI had not hitherto incorporated: 'Bleeding-heart Libertarianism,' based on the work of US intellectuals such as Jeffrey Friedman (1997), who argue that the best arguments for free markets are their actual benefits. That is, markets should be defended and expanded because they provide greater welfare to all, especially the poor, not purely because of a priori reasons. This utilitarian mode of argumentation could allow libertarians to establish broader alliances, as a part of the public is wary of the movements' perceived inattentiveness to poverty and social justice (Bowman, in Scottish Liberty Forum 2013). These ideas are at odds with some of the postulates of classical 'rights-based' libertarians such as Robert Nozick and Ayn Rand, which have gained popularity since the crisis within libertarian circles.

On the other hand, although the Institute has always advocated for expanding civil liberties, this aspect of their work became ever more central. Nowhere is this more conspicuous than on immigration (Bowman 04/2011). The ASI, being libertarian, supports lenient

[22]Charlotte Bowyer is former ASI Head of Digital Policy, working on civil liberties, copyright, and the web.

[23]Ben Southwood is former ASI Head of Policy, interested in monetary regimes and sports economics.

[24]Phillip Salter is former Programmes Director, later Business-features editor at *CityAM* and Director of 'The Entrepreneurs Network.'

[25]Kate Andrews is former ASI Communications Manager and is currently IEA's associate director.

immigration policies, which has at times pitted them against parts of a more unsparing Conservative camp. This could be extended to their views on welfare recipients and the unemployed. Bowman wrote in 2013:

> If you think that unemployment is largely caused by government mismanagement of the economy, it makes no sense to humiliate people for being out of work. If you think that government welfare has crowded out private charity, you shouldn't blame people forced to rely on government disability benefits. (Bowman 06/2013)

This consequentialist approach has been accompanied by a more flexible positioning vis-à-vis their ideological adversaries. Indeed, Bowman has cited as sources of inspiration Converse's (2006) work on belief systems in politics, claiming that the more active and informed political actors are, the least likely they are to modify their views in the face of contravening evidence. 'Bleeding-heart Libertarianism,' in a nutshell, seeks to avoid that fate for libertarians. Concomitantly, Bowman has also collaborated with Bright Blue (Brenton et al. 2014), a conservative think tank that pushes for a more energetic, liberal, and modernised political right that could appeal to the young.

Beyond 2013, it is perhaps curious that the ASI has integrated sociological work on the interplay between diversity and social cohesion (Dobson 04/2015), which is an interesting counterpoint to the Institute's traditionally disdainful relationship with sociology (e.g., Worstall 05/2012). They have also become quite adept at employing and producing internet 'memes' in their social media accounts, seeking to make Adam Smith and pro-capitalist ideas more approachable to a new generation. Public interventions such as these are attempts to engage with those who do not necessarily think of themselves as free-marketeers, in the understanding that "[i]f people in the backs of their minds are thinking [a libertarian] is basically Sarah Palin, or Glenn Beck, then we're in big trouble" (Bowman, in *International Business Times* 2014).

At this juncture, one should highlight intellectual changes in parts of the organisation, which did not necessarily coincide with the broader free-market movement. A case in point is Quantitative Easing (QE), a monetary measure—in the years under consideration, employed in the US and Britain, and much later by the European Central Bank—to inject liquidity. While earlier ASI interventions within our timeframe criticised QE as 'printing money' (Bowman 03/2009), with its associated inflationary risks, others later cautiously supported it (Southwood 02/2013). Indeed, one interviewee said the change of mind came for empirical reasons, as "economies that had implemented QE tend to be in a better shape today" (ASI interview). Nonetheless, this generated differences within the Institute, as "some of the ASI personnel are 'pure Austrians,' meaning they follow the dictates [...] of Ludwig Von Mises. Some [...] are 'empirical Austrians' and follow Friedrich Hayek" (ASI interview). The former are, for theoretical reasons, less supportive of QE, while the latter consider themselves more prone to acknowledging empirical inconsistencies. This outlook mirrored tensions between consequentialists and rights-based libertarians, visible in Libertarian websites (LibertarianHome 2014) and even in the comments section of the ASI blog.

Nevertheless, even if 'Bleeding-heart Libertarianism' has come to inform much of the work written under the Institute's label, changes are neither total nor homogenous. In 2012, the ASI held its first Ayn Rand Annual Lecture, through which it hoped to strengthen its ties with the business and banking world. The invited first speaker was John Allison, then President of CATO. The second was Lars Seier Christensen, ASI fellow and CEO of Saxo Bank. The controversial choice of name for the lectures signalled a push for free markets to be considered "not only an economic argument, but also a principled one" (ASI interview).

In sum, intellectual changes within the ASI, although they certainly exist, are hard to periodise and pin down, and are only visible unevenly across the public interventions done in its name. All in all, in 2014, Pirie argued at the European Students for Liberty (2014) conference in Prague that the crisis had ultimately been good for capitalism, as it reminded us of the need for growth and innovation, and of the urgency of keeping public debt under control. In Pirie's account, this is based on a specific theory of social change based on Darwin and slow evolution rather than on Hegel and dialectic conflict (Pirie 04/2013).

The crisis had, in the end, been another instance of trial and error (Butler 09/2008), which justifies a level of optimism, as the market's self-regulating mechanisms will always tend to better outcomes. To critics of capitalism, Pirie and the ASI would reply that even as Europe and the US muddled through, other economies continued to prosper, especially in the developing world (Lundberg 02/2012). The crisis had even been beneficial for the Institute itself, "as it energised our policy research [and] brought new donors" (ASI interview).

The Muddled and Layered Process of Intellectual Change

I would now like to offer some reflections that sprang from field notes taken during a TNG meeting I attended in 2014. These consist of monthly networking events for libertarians under thirty, who meet for a drinks reception and a brief speech by a distinguished guest. The venue, ASI's mezzanine, was crammed, the audience overwhelmingly young, well dressed, and male, and the address itself was brisk, jovial, and limited to a celebration of the benefits of free trade, in general and in abstract, in the context of the Transatlantic Trade and Investment Partnership negotiations.

The ASI organises many such lectures and social gatherings. Interviewees have described the rationale behind these as an effective way of bringing together people from different domains (particularly free-market academics, young people, politicians, and journalists) in order to coordinate their efforts to exert an influence over policymaking and the public debate. That is, TNG meetings aim to be a node for the organisation and for the spread of libertarian arguments.

But what would justify the regular expense of TNG meetings for a budget-streamlined think tank, given their uncertain and unmeasurable 'return on investment'? At least three discernible reasons are: (a) the ASI accrues part of its funding specifically for their youth activities; (b) TNG allows the Institute to target students and young professionals, fostering a network for collaboration in the medium and long term; and (c) events like these, plus the age of many of its employees, help establish ASI's youthful image, which renews its brand and opens funding avenues.

There might also be more profound reasons. Medvetz (2006: 343), for the case of conservative Wednesday meetings in Washington, argued that such instances function as 'instruments of material power' and 'rituals of symbolic maintenance,' by bringing together conservative advocates through 'relations of force' and 'relations of meaning.' This implies that Wednesday meetings help articulate the US conservative camp by facilitating coordinated teamwork: assigning material and intellectual resources to times, places, and policy areas where they are needed, all while performing boundary work, delimiting those in and out of the movement. Similarly, TNG gatherings—whose format seems less conducive to strategic coordination than Medvetz's case-study—create a propitious atmosphere for young supporters to fulfil similar purposes: to meet, socialise, and help determine what defines their collective. Networking between the young and senior figures helps attendees elucidate what being a free-market 'libertarian' means, generating vigour and cohesiveness in the movement.

The experience conjured some of the underlying questions behind this book: whether and how intellectual change occurs in think tanks after momentous events with high symbolic and material stakes. As many of my colleagues commented in the course of this research, it seemed implausible that one would see many think-tankers proverbially fall on the road to Damascus. This is to be expected. A think tank is in some sense a complex that links ideological positions, networks of people, and institutional arrangements. This complex, which brings together a broad array of actors—funders, experts, politicians, and journalists, young and old—make social relationships inseparable from a shared viewpoint. Indeed, these networks would not exist without a minimal agreement on politics and economics, derived from its history and strengthened by the formation of a canon: Friedman, Hayek, Buchanan, and Smith. And within this complex of links, material and ideological investments are hard to distinguish. In events like TNG meetings, connections, ideas, and resources become, as it were, 'actual,' 'sensible,' and thus reproducible. Think tank events help strengthen 'thought collectives' (Mirowski 2013).

Nevertheless, this is not to say members of such networks are unable to change their mind, nor that reflexive and highly educated think-tankers, however partisan, would have no inclination to examine their views, especially after an event as overriding as the financial crash. It is only to say that ideas, regardless of their perceived accuracy, over time structure social networks. Besides, as interviewees have stressed, there are relatively few procedural controls over what is said under the ASI label, presumably the red line being, *ceteris paribus*, a preference for free markets and individual choice over government action.

Indeed, views within the ASI can shift, one exemplary case being its stance on QE. These illustrations of intellectual change signal rifts that can separate members' opinions, and which amount to tensions within its camp that often go unnoticed in agonistic views of the public debate. These include: differences between consequentialist and principled arguments for capitalism[26]; between 'mainstream' and 'radical' monetary policy; between downplaying the economic crisis or tackling it head-on; and between 'pure' and 'empirical' Austrians. And although these frictions can be mitigated or spread across time, ASI's standing benefits from a degree of mutability, as long as it is not seen as making it go adrift—a criticism they have received from other libertarians (e.g., LibertarianHome 26/11/2014). It could be claimed that within the free-market 'thought collective' as within any other, there are 'sacred' and 'profane' issues. And, when dealing with profane aspects of policy and when seeking to reach larger publics, it is perhaps better to seem pragmatic—as opportunist *bricoleurs* rather than as 'paradigm men' (Carsensten 2011). Those efforts are visible in the format and content of ASI's public interventions: in a greater variability of narratives in non-core matters such as QE, and in a less confrontational approach to issues such as poverty and welfare.

[26]Pirie's writings are awash with principled reasons for free markets, which is likely to continue parallel to consequentialist arguments. See, for instance, this excerpt: "privatization was not a series of policies but a system that fitted into a theoretical approach" (Pirie 2012a: 79).

Most academic accounts of neoliberal thought after the crisis, to my knowledge, have been denunciatory of the doggedness of its defenders, only rarely acknowledging internal frictions. Some have argued, not without reason, that neoliberal thought has proven resilient partly because it counts with rhetorical devices that allow their ideas to remain distant from their actual implementation (Schmidt and Thatcher 2013). This dissociation is achieved through, for instance, blaming 'crony-capitalism' instead of capitalism itself, central banks instead of bankers, and overzealous regulators instead of socially irresponsible private actors. Critics of neoliberalism, among which I count myself, often accuse this type of argument of being an instance of confirmation bias. Concerning the fact that many neoliberal thinkers have not revised their beliefs after 2008, and on the contrary have emerged more resolute, Philip Mirowski (2013: 120) cited *When Prophecy Fails* (Festinger et al. 2008).[27] An almost infinite number of auxiliary hypotheses can be put forward to counter an argument that challenges the mainspring of a collective's self-definition. And if such collective is to survive—especially when there are important sources of external support—it will react with newfound resolve. Against the motion '*Was Karl Marx right?*' at a Cambridge Union debate, Pirie maintained that:

> Capitalism certainly faced a crisis in 2008, but it is still with us, as yet uncollapsed. It is evolving and responding to the changes that are needed and, as before, when the dust of crisis has settled, it will be a new version of capitalism that goes on to generate more wealth and to expand the opportunities open to humankind. (Pirie 04/2013)

It would be easy to criticise such ideas for their inability to be disproven. But while this argument has merit, one should be aware that such a case can only be done by those outside the network of social relationships whose core is the idea that free markets are always best.

[27]In this social psychology classic, Festinger explores the coping strategies of a sect gathered to wait for the apocalypse in the hours after it fails to materialise. Festinger finds that, when confronted with evidence that undermines a group's core beliefs, members tend to devise ever more elaborate ways of rationalising conflicting evidence.

Furthermore, part of the definition of the 'sacred' is, after all, its ambiguity and malleability.[28] Free markets are never fully realised in this world, and their principles are open to negotiation in changing circumstances. Or, in more prosaic terms, the contours of a collective are always in need of coordination and adaptation. Hence why parts of the ASI have come to challenge some of the tenets of rights-based libertarian thought; particularly, I submit, when they are seen as detrimental to their attempts to garner respect and authority across larger publics.

Furthermore, the informational, contested and highly technical character of the 2008 crisis renders difficult any attempt to convince those with ideological stakes in the game, especially as it occurred in tandem to a generalised discrediting of economic experts. For a think tank with a history of antagonistic engagement in the policy debate, an abrupt repositioning would most likely have been damning, severing vital intellectual and institutional ties.[29] However, the ASI could not be entirely led by *hysteresis* either. Going forward, this tension will become noticeable once a new generation of libertarians, less encumbered by the Cold War, attempt to convince those who are not yet convinced of the case for free markets—or, perhaps, if and when they clash against another young subsection of the libertarian movement that is more enamoured with Ayn Rand and much more closely aligned with a Darwinian view of economics.

The internal discrepancies described above raise questions over the level in which intellectual change tends to occur. Instances of repositioning are visible in public interventions on the link between QE and inflation, but not in the inviolability of private property. Repositioning is a sign of flexibility, but this can be shown in some issues and not in others and can be useful for reaching some publics rather than others—particularly those not in the 'thought collective' but who could be brought to the fold on specific issues. All the while, in their old

[28]Durkheim already knew well of the ambiguity of the sacred: "[w]hile the fundamental process is always the same, different circumstances color it differently" (Durkheim 1995 [1912]: 417).

[29]An example of this is when the neoliberal think tank network 'Stockholm network' argued for state support for the pharmaceutical industry, which prompted many think tanks to disaffiliate themselves (*Daily Telegraph* 2009a).

audiences, tensions brew under the surface. Thus, a researcher attempting to unravel the crisis narratives of the ASI would not achieve much by simply colliding one 'policy paradigm' against another, claiming one is wrong and another correct. My objective in this book is, on the contrary, to compare these ways of thinking and see how they change. In ASI's case, this meant avoiding considering them immediately as ideologues, as is often the case in the literature, but allowing room to observe renewal where it happens. At this point, I posit that instances of repositioning, as both the experience of NEF and ASI suggests, were particularly visible in the format of their public interventions (especially concerning their self-presentation), which are linked to efforts to expand their appeal. One interviewee commented, in a manner that few would associate with the ASI:

> We try to avoid presenting our philosophical views as a single cohesive unit that you either have to believe or disagree with. [We try] to make our arguments based on evidence and make them in such a way that a non-free marketer can find them useful and possibly […] persuasive without having to stop and be a free-marketer. The reason for that is because I think that a lot of British politics is like an echo chamber and I think think tanks have a privileged position in that we are almost the only people in the country who can actually say what we think. (ASI interview)

References

Baert, P., & Morgan, M. (2015). *Conflict in the academy: A study in the sociology of intellectuals*. London: Palgrave Pivot.

BBC. (2007). *Northern Rock: Expert views*. Accessed 20 March 2015. http://news.bbc.co.uk/1/hi/business/6999246.stm.

BBC. (2013). *Eamonn on BBC Breakfast TV on fizzy drinks tax*. Accessed 21 March 2015. https://www.youtube.com/watch?v=YWIyjUNlZ08.

Brenton, M., Maltby, K., & Shorthouse, R. (Eds.). (2014). *The moderniser's manifesto*. London: Bright Blue.

Butler, E. (2007). *Adam Smith: A primer*. London: IEA.

Butler, E. (2008). *The best book on the market: How to stop worrying and love the free economy*. Oxford: Capstone.

Butler, E. (2009a). *The rotten state of Britain: Who is causing the crisis and how to solve it*. London: Gibson Square.
Butler, E. (2009b). The financial crisis: Blame governments, not bankers. In P. Booth (Ed.), *Verdict on the crash: Causes and policy implications* (pp. 51–58). London: IEA.
Butler, E. (2010a). *The alternative manifesto: A 12-step programme to remake Britain*. London: Gibson Square.
Butler, E. (2010b). *Ludwig Von Mises: A primer*. London: IEA.
Butler, E. (2010c). *Austrian economics: A primer*. London: IEA.
Butler, E. (2011a). *Milton Friedman: A concise guide to the ideas and influence of the free-market economist*. Petersfield: Harriman House.
Butler, E. (2011b). *The condensed wealth of nations*. London: Adam Smith Institute.
Butler, E. (2012a). *Friedrich Hayek: The ideas and influence of the libertarian economist*. Petersfield: Harriman House.
Butler, E. (2012b). *Public choice: A primer*. London: IEA.
Butler, E. (2013). *Foundations of a free society*. London: IEA.
Carstensen, M. (2011). Paradigm man vs. the bricoleur: Bricolage as an alternative vision of agency in ideational change. *European Political Science Review, 3*(1), 147–167.
Clark, S. (2012). *They can't help themselves: Anyone who contradicts the tobacco control industry must be a stooge of Big Tobacco*. London: Forest. Accessed 15 March 2015. http://taking-liberties.squarespace.com/blog/2012/2/20/asi-acting-as-mouthpiece-for-the-tobacco-industry-says-ash.html.
Cockett, R. (1995). *Thinking the unthinkable: Think tanks and the economic counter-revolution 1931–1983*. London: HarperCollins.
Converse, P. (2006 [1964]). The nature of belief systems in mass publics. *Critical Review: A Journal of Politics and Society, 18*(1), 1–74.
Daily Telegraph. (2009a). Free-market network demands bail-out for pharmaceutical industry. Accessed 20 February 2015. http://blogs.telegraph.co.uk/news/alexsingleton/8145947/freemarket_network_demands_bailout_for_pharmaceutical_industry/.
Daily Telegraph. (2009b). Madsen Pirie: The Archbishop of Canterbury caricatures consumers and fires at token targets. Accessed 18 February 2015. http://www.telegraph.co.uk/comment/personal-view/6315892/The-Archbishop-of-Canterbury-caricatures-consumers-and-fires-at-token-targets.html.

Daily Telegraph. (2009c). Madsen Pirie: Lord Stern is wrong—Giving up meat is no way to save the planet. Accessed 20 February 2015. http://www.telegraph.co.uk/news/earth/environment/climatechange/6445930/Lord-Stern-is-wrong-giving-up-meat-is-no-way-to-save-the-planet.html.

Daily Telegraph. (2012). It doesn't matter who funds think tanks, but if it did, Left-wing ones would do particularly badly. Accessed 20 February 2015. http://blogs.telegraph.co.uk/finance/timworstall/100018107/it-doesnt-matter-who-funds-think-tanks-but-if-it-did-left-wing-ones-would-do-particularly-badly/.

Dean, M. (2012). Rethinking neoliberalism. *Journal of Sociology, 50*(2), 150–163.

Denham, A., & Garnett, M. (1998). *British think tanks and the climate of opinion*. London: UCL Press.

Desai, R. (1994). Second hand dealers in ideas: Think tanks and Thatcherite hegemony. *New Left Review, 203*(1), 27–64.

Djelic, M. L. (2014). *Spreading ideas to change the world: Inventing and institutionalizing the neoliberal think tank*. Social Science Research Network. Accessed 22 February 2015. http://papers.ssrn.com/sol3/papers.cfm?abstract_id=2492010.

Durkheim, E. (1995 [1912]). *The elementary forms of religious life*. New York: The Free Press.

DWP. (2010). *Consultation responses to 21st century welfare*. Accessed 25 March 2015. https://www.gov.uk/government/uploads/system/uploads/attachment_data/file/181144/21st-century-welfare-response.pdf.

European Students for Liberty. (2014). *Madsen Pirie: Prospects for liberty*. Accessed 25 March 2015. https://www.youtube.com/watch?v=BHP0GrBhuNs.

Festinger, L., Riecken, H., & Schachter, S. (2008 [1956]). *When prophecy fails: A social and psychological study of a modern group that predicted the destruction of the world*. London: Pinter & Martin.

Friedman, J. (1997). What's wrong with libertarianism. *Critical Review, 11*(3), 403–467.

Guardian. (2008a). *Will the market bounce back soon?* Accessed 16 February 2015. http://www.theguardian.com/money/2008/nov/01/moneyinvestments-investmentfunds?INTCMP=SRCH.

Guardian. (2008b). 167 *Eamonn Butler: Sentamu and the city*. Accessed 18 February 2015.

Guardian. (2012). *James Meadway: Development's fat cats have been gorging on private sector values*. Accessed 20 March 2015. http://www.theguardian.com/commentisfree/2012/sep/19/development-fat-cats-private-sector-values.

Heffernan, R. (1996). Blueprint for a revolution? The politics of the Adam Smith Institute. *Contemporary British History, 10*(1), 73–87.
International Business Times. (2014). Rise of the new libertarians: Meet Britain's next political generation. Accessed 18 February 2015. http://www.ibtimes.co.uk/rise-new-libertarians-meet-britains-next-political-generation-1469233.
Jackson, B. (2012). The think tank archipelago: Thatcherism and neoliberalism. In B. Jackson & R. Saunders (Eds.), *Making Thatcher's Britain* (pp. 43–61). Cambridge: Cambridge University Press.
Jacobs, R., & Townsley, E. (2011). *The space of opinion: Media intellectuals and the public sphere*. Oxford: Oxford University Press.
Jacques, P., Dunlap, R., & Freeman, M. (2008). The organisation of denial: Conservative think tanks and environmental scepticism. *Environmental Politics, 17*(3), 349–385.
James, S. (1993). The idea brokers: The impact of think tanks on British government. *Public Administration, 71,* 491–506.
Kay, L., Smith, K., & Torres, J. (2013). Think tanks as research mediators? Case studies from public health. *Evidence and Policy, 59*(3), 371–390.
LibertarianHome. (2014). *Sam Bowman promoted to deputy director of the Adam Smith Institute*. Accessed 27 March 2015. http://libertarianhome.co.uk/2014/11/sam-bowman-promoted-to-deputy-director-of-the-adam-smith-institute/.
LSE. (2011). *Keynes v Hayek*. Accessed 20 March 2015. http://www.lse.ac.uk/newsAndMedia/videoAndAudio/channels/publicLecturesAndEvents/player.aspx?id=1107.
McGoey, L. (2012). Strategic unknowns: Towards a sociology of ignorance. *Economy & Society, 41*(1), 1–16.
Medvetz, T. (2006). The strength of weekly ties: Relations of material and symbolic exchange in the conservative movement. *Politics & Society, 34*(3), 343–368.
Medvetz, T. (2012). *Think tanks in America*. Chicago: University of Chicago Press.
Mirowski, P. (2013). *Never let a serious crisis go to waste*. London: Verso.
Pautz, H. (2012). The think tanks behind 'cameronism'. *British Journal of Politics and International Relations, 15*(3), 362–377.
Pirie, M. (1988). *Micropolitics: The creation of successful policy*. Aldershot: Wildwood house.
Pirie, M. (2007). *How to win every argument: The use and abuse of logic*. London: Continuum.
Pirie, M. (2008). *Freedom 101*. London: Adam Smith Institute.

Pirie, M. (2009). *Zero base policy*. London: Adam Smith Institute.
Pirie, M. (2012a). *Think tank: The history of the Adam Smith Institute*. London: Biteback.
Pirie, M. (2012b). *Economics made simple: How money, trade and markets really work*. Petersfield: Harriman House.
Pirie, I. (2012c). Representations of economic crisis in contemporary Britain. *British Politics, 7*(4), 341–364.
Sapiro, G. (2009). Modèles d'intervention politique des intellectuels: Le cas français. *Actes de la recherche en sciences sociales, 176–177*(1–2), 8–31.
Schlichter, D. (2011). *Paper money collapse: The folly of elastic money and the coming monetary breakdown*. London: Wiley.
Schmidt, V., & Thatcher, M. (2013). *Resilient liberalism in Europe's political economy*. Cambridge: Cambridge University Press.
Scottish Liberty Forum. (2013). *Sam Bowman: The free market road to social justice.* Accessed 30 March 2016. https://www.youtube.com/watch?v=Rptm0yyvPik.
Seiler, C., & Wohlrabe, K. (2010). A critique of the 2009 global "go-to think tank" ranking. *CESifo DICE Report, 8*(2), 60–63.
Selgin, G. (1994). Are banking crises free-market phenomena? *Critical Review: A Journal of Politics and Society, 8*(4), 591–608.
Stone, D. (2007). Recycling bins, garbage cans or think tanks? Three myths regarding policy analysis institutes. *Public Administration, 85*(2), 259–278.
Tax Justice Network. (2010). *Adam Smith, the Adam Smith Institute, and flat taxes.* Accessed 16 February 2015. http://taxjustice.blogspot.co.uk/2010/01/adam-smith-adam-smith-institute-and.html.
Trevisan, J. (2012). Global go-to 2011 think tank ranking: An analysis. *International Center for Climate Governance*. Accessed 25 February 2015. http://www.iccgov.org/FilePagineStatiche/Files/Publications/Reflections/03_reflection_february_2012.pdf.
Whyte, J. (2013). *Quack policy*. London: IEA.
Worstall, T. (2010). *Chasing rainbows: How the green agenda defeats its aims.* London: Stacey International.

Think Tank Reports and Blog Posts (ASI, available at adamsmith.org)

Ambler, T. (11/2008). *The financial crisis: Is regulation cure or cause?*
Ambler, T., & Boyfield, K. (11/2007). *Stemming the growth of UK regulatory agencies.*

Ambler, T., & Saltiel, M. (09/2011). *Bank regulation: Can we trust the Vickers report?*
ASI. (01/2010). *2009 annual review.*
ASI. (05/2010). *The effect of capital gains tax rises on revenues.*
ASI. (01/2011). *2010/2011 annual review.*
ASI. (01/2012). *2011 annual review.*
Bowman, S. (04/2011). *Immigration restrictions make us poorer.*
Bowman, S. (06/2013). *Don't hate the players, hate the game.*
Bowman, T. (03/2009). *On quantitative easing.*
Boyfield, K. (03/2009). *The Turner review: A case of poacher turned gamekeeper?*
Boyfield, K. (10/2009). *Cure or disease? The unintended consequences of regulation.*
Butler, E. (09/2008). *A circle in a spiral.*
Butler, E. (10/2008). *Don't knock the system: Politics caused this crisis of capitalism.*
Butler, E. (03/2009a). *A Labour-made crisis.*
Butler, E. (03/2009b). *Believers in free-market are fighting back.*
Butler, E. (04/2009). *Save the tax havens—We need them.*
Butler, E. (06/2009). *Government debt: That'll be £2.2 trillion, please.*
Butler, E. (06/2010). *Reboting government.*
Butler, E. (06/2011). *Economists? What economists?*
Butler, E. (05/2012). *The rotten state of our democracy.*
Butler, E., & Teather, R. (04/2009). *Parliamentary fatcats.*
Clougherty, T. (09/2008). *The financial crisis in bullet-points.*
Clougherty, T. (01/2009). *How to promote the free market in 2009.*
Croft, J. (04/2011). *Profit-making free schools: Unlocking the potential of England's proprietorial schools sector.*
Dobson, J. (04/2015). *The ties that bind: Analysing the relationship between social cohesion, diversity, and immigration.*
Graham, D. (08/2010). *Global player or subsidy junky? Decision time for the BBC.*
Hansen, F. (02/2008). *It's government intervention, stupid!*
Hawkins, N. (06/2010). *The party is over: A blueprint for fiscal stability.*
Hawkins, N. (10/2010). *Privatization revisited.*
Hill, H. (02/2012). *What turns doctors into tyrants?*
Lal, D. (07/2009). *The great crash of 2008: Are governments or markets to blame?*
Lundberg, J. (02/2012). *The triumph of global capitalism.*
Mchangama, J. (06/2009). *The war on capitalism.*

Oliver, H. (01/2012). *Why MigrationWatch is wrong about immigration and unemployment.*
Patterson, R. (03/2009). *A brief history of the social rights myth.*
Pirie, M. (08/2009). *It is no time for Westminster to be squeamish over spending cuts.*
Pirie, M. (10/2009). *How David Cameron can reverse Labour's unjustified attacks on civil liberties.*
Pirie, M. (04/2013). *Why Marx was wrong about capitalism.*
Rawcliffe, D. (03/2010). *Arts funding: A new approach.*
Redwood, J. (10/2009). *Credit crunch: The anatomy of a crisis.*
Saltiel, M. (03/2009). *What went wrong? An agenda for the G20.*
Saltiel, M. (12/2010). *On borrowed time: Avoiding fiscal catastrophe by transforming the state's intergenerational responsibilities.*
Saltiel, M. (09/2011). *A botched opportunity: Why the Vickers report won't fix the financial sector.*
Saltiel, M., & Young, P. (10/2011). *The revenue and growth effects of Britain's high personal taxes.*
Senior, I. (11/2002). *Consigned to oblivion.*
Sidwell, M. (02/2008). *Unfair trade.*
Simpson, D. (06/2009). *The recession: Causes and cures.*
Skousen, M. (02/2009). *Has Keynes trumped Adam Smith?*
Southwood, B. (07/2013). *Despite its problems, QE might be right.*
Stanfield, J. (03/2010). *The broken university.*
Wellings, R. (Ed.). (12/2009). *Beginner's guide to liberty.* London: ASI.
Whig. (11/2011). *Do we have a capitalist economy?*
Worstall, T. (07/2010). *Is it because they iz doctors?*
Worstall, T. (05/2012). *Sociologists doing economics.*
Worstall, T. (09/2012). *Can we please kill the idea that Adair Turner or Robert Skidelsky are economists?*

5

The National Institute of Economic and Social Research: The Shifting Fortunes of Expert Arbiters

The National Institute of Economic and Social Research (NIESR), established in 1938, is one of the oldest policy research institutes in Britain. Launched by initiative of the then Bank of England (BoE) Director Baron Josiah Stamp—and supported by funds from the Rockefeller Foundation and the Leverhulme, Halley Stewart, and Pilgrim Trusts (Denham and Garnett 1998: 57–59)—NIESR's mission was to provide rigorous independent analysis to counterbalance the monopoly the Treasury had on publishing data on the British economy (Jones 05/1988; 04/1998).

Since its foundation, the Institute benefitted from close links to academia and public service, including William Beveridge from the LSE (NIESR's first Chairman), Henry Clay from the BoE (first Deputy Chairman) and Noel Hall from the University of Oxford (first Director). Many of these economists had coincided, some years earlier, at the Economic Advisory Council organised by the Ramsay

Throughout this chapter, references in the format (MM/YYYY) include those of this think tank's own academic journal, the *National Institute Economic Review*.

M. González Hernando, *British Think Tanks After the 2008 Global Financial Crisis*, Palgrave Studies in Science, Knowledge and Policy,
https://doi.org/10.1007/978-3-030-20370-2_5

MacDonald premiership to present the government with expert advice that could inform its response to the Great Depression—an instance which, with its internal quarrels between 'Keynesians' and 'Free Traders,' was an apt synecdoche of the economic debates of the day, and indeed of those to come. Accordingly, much of NIESR's senior staff has since then been closely connected to (and sometimes actually been) some of the most influential British economists of their time, perhaps especially those working for public institutions such as the BoE, the Treasury, the Cabinet Office, and other government departments.[1]

The Institute's founding coincides with a growing preponderance of quantitative tools and models in the economics discipline, especially since the 1930s. At that time, these methodologies informed a wider effort to supply the economic and demographic knowledge governments demanded in the aftermath of the 1929 crisis and the Second World War, so as to produce socio-scientific evidence to justify governance decisions (Mirowski 2002). In the years leading to NIESR's foundation, economists increasingly sought to provide authoritative knowledge that policymakers could refer to, specifying a scientifically sound course of action under a broadly positivistic matrix. More specifically still, and given its origins in British economics, NIESR's foundation is linked to a then dominant post-war technocratic ethos, with a notable focus on growth and full employment—hence its emphasis on GDP, productivity, labour, and industry. However, notwithstanding continuous efforts to be perceived as ideologically neutral, after the 1970s monetarist apogee (and a political reframing of Keynes' ideas) NIESR has many times been called 'Keynesian' by those contesting their claims (e.g., ConservativeHome 2013).

Internationally, the Institute is comparable to the National Bureau of Economic Research (NBER) in the United States and the *Konjunktur* institutes of northern Europe.[2] These organisations, much like NIESR, produce independent forecasts and indicators to supply governments with policy-relevant econometric knowledge. They also share NIESR's

[1]Although NIESR went through a hiatus during WWII, due to the concentration of economists in the war effort, it became stronger after the war (Jones, op. cit.; Robinson 05/1988).

[2]These include the *Deutsches Institut für Wirtschaftsforschung* in Germany and the Swedish *Konjunkturinstitutet*, founded in 1925 and 1937, respectively.

emphasis on political neutrality, strong academic credentials, and an insistence on "the scientific character of their work" (Reichman 2011: 567). NIESR is thus one of the institutional embodiments of the aspiration of economists in the second half of the twentieth century to render their discipline a science of governance and of the desire of policymakers for the existence of such a science.

Nevertheless, comparable organisations elsewhere rely either on large charitable endowments (NBER) or direct government support (*Konjunktur* institutes). In the absence of both sources, and partly to supplement a dearth of substantial core funding, NIESR has complemented its macroeconomic work with short-term research contracts with government departments and other organisations on an array of economic and social issues. In the 1960s, NIESR and the Treasury had agreed it would be improper for the institute to receive direct government funds as it "might raise doubts about its independence" (Jones 05/1988: 44). In addition, NIESR is the only think tank in this study that receives funding from academic research councils (Jones 05/1988). Given their research profile, NIESR has tended to favour applied economics research in policy-relevant areas not commonly covered by university departments. A substantial proportion of NIESR's academic grants have come, from 1965 onwards, from the Social Science Research Council (SSRC), later Economic and Social Research Council (ESRC).

Due to NIESR's position at the crossroads of academia and government, its list of former directors and presidents includes eminences in both (and in between) those fields. NIESR's first generation was succeeded by future BoE Executive Director Christopher Dow,[3] the Institute's first Deputy Director, and by subsequent Directors Bryan Hopkin (1952–1957),[4] Christopher Saunders (1957–1965),[5]

---

[3]Christopher Dow was also former Assistant Secretary General of the OECD and Fellow of the British Academy (*Independent* 1998).

[4]Bryan Hopkin was chair at the University of Cardiff and, among many other public service positions, head of the Government Economic Service and Chief Economic Advisor to the Treasury during Healey's premiership (see *Daily Telegraph* 2009).

[5]Before joining NIESR, Christopher Saunders was Deputy Director of the Central Statistical Office, and later became Director of Research for the UN Economic Commission for Europe (Frowen 1983).

David Worswick (1965–1982),[6] Andrew Britton (1982–1995) and, within our timeframe, Martin Weale (1995–2010). After Weale left in 2010 to join the BoE Macroeconomic Policy Committee (MPC)—which decides over the UK base interest rate—he was succeeded by Jonathan Portes (2011–2015), a former senior civil servant who had worked under Conservative and Labour governments.

To give another example of NIESR's centrality across networks of expert governance: two of its former directors (Andrew Britton and Martin Weale) were part of the Treasury's Panel of Independent Forecasters established in 1992 following Britain's exit from the European Exchange Rate Mechanism (Budd 1999). After said panel was disbanded and replaced by the BoE's MPC, NIESR staff continued to be loosely related to its heir institution (Evans 1999: 190). What is more, current and former members of the MPC have also been part of NIESR or its Council of Management—e.g., Charles Goodhart, Kate Barker, Charles Bean, Alan Budd, Tim Besley, and Weale himself. For those reasons, it is impossible to disentangle NIESR's history from that of British economic policymaking, or indeed, of British policy-oriented economics (Coats 2000: 32). It is under that light that NIESR's strive towards rigour and non-partisanship should be observed, being a historically privileged 'conduit' between academia and the state.

However, as a result of the end of the post-war consensus, Denham and Garnett (1998) suggested NIESR's halcyon years are long past, especially after their public discords with the Thatcher government, the rise of more media-oriented advocacy think tanks, and a growing mistrust among sectors of the public of academic economics. NIESR, nevertheless, maintains a central position as expert arbiters: its research continues to be contracted and its approval prized by all parties, and its forecasts provide a privileged vantage point to assess economic performance. UPenn's *Global go to* think tank ranking lists NIESR 60th among the top domestic economic policy think tanks in 2012, and 62nd in the three following years (McGann 2013–2016).[7]

[6] Possibly NIESR's first mainly academic director, David Worswick was based at the University of Oxford Institute of Statistics.

[7] See, for comparison, ASI top-ten positions in the same categories (Chapter 4).

Organisational and Funding Structure

Since 1942, NIESR's headquarters are located in 2 Dean Trench Street, a Georgian building in a tranquil area of Westminster, a few blocks south of Parliament and close to government departments, international organisations, and other think tanks. The premises accommodate an in-house library and regularly hosts small events—most commonly press briefings, training sessions, seminars, and talks. Variably over forty members of staff work for NIESR: a small administration and communications team, twenty to thirty resident research staff, plus visiting fellows (mainly senior academics from Britain and overseas).

NIESR's research has traditionally been structured in three broad areas: macroeconomic modelling and forecasting; education, training, and employment; and international economics. Under these categories, NIESR undertakes mostly quantitative (but also some qualitative) research within several themes. These have included ageing, migration, productivity, pensions, welfare, health, and wellbeing, employment policy, and innovation. These themes have varied, swayed by the policy climates and debates of the time. In their words, "[NIESR's] research interests are constantly changing in response to new needs but embrace most of the issues that shape economic performance."[8] In terms of format, NIESR's public interventions most commonly have taken the form of working papers, economic indicators, press releases, commissioned research reports, and peer-reviewed articles.

The majority of NIESR's work qualifies as applied economics, although social scientists from other disciplines are also employed, notably demographers, sociologists with a strong quantitative profile, and those with expertise in areas broadly linked to employment, productivity, and growth (e.g., labour, migration, skills, innovation, industrial relations, education). NIESR's loose connections to University departments—both through shared staff and research—include the LSE, Brunel, UCL, Queen Mary, Melbourne, and many others. Although

[8]Accessed 10 October 2015, https://web.archive.org/web/20101219124903/http://www.niesr.ac.uk/aboutniesr2.0.php.

none of these associations is permanent,[9] they often involve collaborating in specific research projects, among which one can mention the ESRC-funded Centre for Economic Performance (CEP), Centre for Learning and Life Chances in Knowledge Economies and Societies (LLAKES), and Centre for Macroeconomics (CFM). Further, much of NIESR's staff is affiliated with comparable think tanks and academic institutions overseas—e.g., IZA and CESifo in Germany, WIIW in Austria, ESRI in Ireland—and international organisations—notably the IMF and OECD. NIESR is also part of EUROFRAME, a network of independent economic research institutes from across Europe producing analysis and forecasts on European economies.

Being NIESR a charity, the work of its permanent staff is overseen by a board of trustees, the Institute's Council of Management. It is mainly composed of distinguished members of the economics profession, as well as businesspeople and politicians from across the spectrum. To name but a few, between 2007 and 2013 they have included Lord Burns,[10] Tim Besley,[11] Diane Coyle,[12] Heather Joshi,[13] John Hills,[14] Bronwyn Curtis,[15] Peter Kellner,[16] John Llewelyn,[17] as well as former Labour Chancellor Alastair Darling, Labour MP Frank Field and Conservative MP Jesse Norman. To these, one could add over 170 governors, chosen 'to extend [NIESR's] outreach and impact.'

[9]By way of contrast, the Policy Studies Institute—formerly Political and Economic Planning—a historically comparable think tank that also has a strong academic profile, depends nowadays on the University of Westminster.

[10]Lord Burns GCB is a life peer, former Treasury Chief Economic Advisor and former Chairman of Abbey plc (Santander bank subsidiary).

[11]Tim Besley CBE is Professor of Economics at the LSE, former member of the BoE MPC, and former president of the International Economic Association.

[12]Diane Coyle OBE is a Professor at the University of Manchester, former advisor to the Treasury, and former Vice-chairman of the BBC, and current NIESR Council Chair.

[13]Heather Joshi CBE is Emeritus Professor of Economic and Developmental Demography at the University of London. She is also the former Director of the Centre for Longitudinal Studies and of the UK Millennium Cohort Study.

[14]John Hills is Head of Centre for the Analysis of Social Exclusion (CASE), LSE.

[15]Bronwyn Curtis OBE is former HSBC Head of Global Research and current Non-Executive Director of JP Morgan Asian Investment Trust.

[16]Peter Kellner is a political commentator and Head of the polling company YouGov.

[17]John Llewelyn is Founding Partner at Llewellyn Consulting, former Managing Director at Lehman Brothers and adviser to HM Treasury.

Given the above, NIESR's audiences and networks are presumably more specialised than those of advocacy-oriented think tanks and include comparable institutes overseas, academics, and experts in the BoE, the Treasury and government departments—that is, the *epistemic elite* of economic policymaking. Consequently, although the media regularly covers the Institute's work—notably their forecasts and their public interventions as arbiters of policy—NIESR has traditionally held a relative distance from nonspecialised political disputes, especially where they involve normative judgments. Partial exceptions to this rule have been when NIESR deliberates on whether the economic evidence supports a policy agenda, as well as a prioritisation of economic growth and productivity over other economic variables (e.g., short-term deficits) or political considerations (e.g., anti-immigration sentiment).

Perhaps because of its predominantly specialist audiences, NIESR has historically had a small communications team. NIESR's blog was only created in January 2012 and its social media presence is arguably underdeveloped when compared to advocacy-oriented organisations. An institutional Twitter account was also only opened in 2012. Nevertheless, NIESR's forecasts receive considerable media attention, especially when the short-term future of the economy is uncertain. Back in the 1960s, when the Institute was one of the few independent economic forecasters in Britain, NIESR had to prevent their data being leaked as they could affect stock markets (Jones, op. cit.).

Regarding funding, the Institute's finances are transparent: its sources are publicly available in relevant Charity Commission documents and are also often mentioned in their reports. NIESR is supported by a variety of patrons: public, private, and charitable. Historically, the Institute has received most of its funds in the modality of research contracts, allocated to specific teams, often through grants to resident and visiting scholars, much like a "university department" would (NIESR interview). Hence, besides subscriptions to their monthly GDP forecasts and publications and its Corporate Membership Scheme, NIESR receives few core donations. The closest likeness in the UK, both in its academic profile and its public role as policy arbiter, is the IFS, albeit with a different financial structure, since a sizeable proportion of IFS researchers are doctoral candidates or have a University salary (NIESR interview).

NIESR has historically depended on public sources, such as government departments (HM Treasury; DWP; BIS; DfID), public institutions (ONS, BoE, NAO), and academic funding councils (ESRC). The Institute has also enjoyed support from local government and EU agencies, including the Scottish Government, the Welsh Development Agency, Eurostat, and the European Commission. To these one should add at least four other types: charitable foundations and grant-making bodies (e.g., NESTA, Leverhulme Trust, Nuffield Trust, JRF); other think tanks, charities, and sectoral organisations (e.g., HECSU, ODI, TUC); funding from the private sector (e.g., Abbey plc, Barclays, Ernst and Young, National Grid, Marks and Spencer, Unilever, Rio Tinto); and comparable overseas sources (e.g., Norwegian Research Council, Nomura Research Institute, Sveriges Riksbank, IZA).

Subscriptions to the Institute's products are another source of income, the foremost being the National Institute Global Econometric Model (NiGEM). Launched in 1969, NiGEM is a demand-side based computer model built to prognose economic performance and probing how varying economic factors might affect it (Jones 04/1998; Evans 1999: 19–20). Successive researchers have hewn this model to render it ever more sophisticated and responsive, and is capable of running scenarios—e.g., oil price instability, sovereign debt crisis, sustained unemployment—to forecast possible developments in the global economy. One of NiGEM's key advantages is that its predictions acknowledge that "people's expectations are forward-looking, and based on what [they] predict will happen instead of being the average of past experience" (NIESR 08/2007: 2). This model underpins the Institute's forecasts and enjoys a broad base of external subscribers (including the HM Treasury, BoE, ECB, IMF, and the OECD). As such, NiGEM is among the most important of NIESR's resources, providing it with prestige, income, and a tool to intervene consistently in the policy debate. However, because of tighter budgets after 2008, many clients cancelled their subscription (NIESR, 08/2009).

Additionally, since 1959, the Institute publishes the *National Institute Economic Review* (NIER), an academic journal that, in the words of an interviewee: "[tries to] have policy relevance [and that] because of [its] relatively short turnaround can pick themes that are relevant to what's going on in the real world" (NIESR interview). Somewhere in between a scholarly journal and a policy-oriented quarterly, NIER customarily includes NiGEM-based economic forecasts and assessments of the prospect of the UK and global economy. The journal publishes research by both NIESR staff and outside academics, either by open submission, direct invitation, or calls for papers—often by authors loosely linked to the Institute through some of the organisations mentioned above—e.g., those at the CEP (Bagaria et al. 2012), the IMF (Babecký et al. 2008), and Council members (Budd 04/2010). NIER issues are often thematic and include research on topical areas in the policy debate (e.g., deficits, banking regulation, migration, the Eurozone crisis).

The Institute's funding figures, as reported to the Charity Commission, are displayed (Table 5.1). A significant and growing proportion of NIESR's income is from fixed-term research projects, while sales of NiGEM and publications provide some proceeds, if involving running costs. Financially, NIESR was exposed to considerable risks linked to the crisis—through diminished funding from public sources following austerity and cost-cutting from subscribers. As a result, the Institute had a tight or negative yearly balance, aggravated by the fact that "government contracts often don't really cover the costs [of research]" (NIESR interview).

Although savings have allowed NIESR some room to weather funding shortfalls, partly as a result of financial pressures, the preponderance of different types of sponsors has changed considerably. NIESR's accounts also provide information on its sources of research funding, summarised below (Fig. 5.1), which displays a noteworthy increase in the proportion of charitable and 'other' (mostly private) funding, from meagre levels in 2007 to almost half in 2013, while public donors—e.g., ESRC and Government departments—have decreased in relative importance.

In addition to this shifting financial climate, NIESR faced staffing challenges due to difficulties in retaining researchers, the time required to train in-house experts, and the paucity of external candidates with the necessary skills (NIESR 03/2012: 2). These are partly due to the Institute's competition with academic institutions, and to the fact that ESRC terms tend to prioritise hiring fixed-term junior (rather than

Table 5.1 NIESR financial overview

NIESR budget (£)	2007–2008	2008–2009	2009–2010	2010–2011	2011–2012	2012–2013
Research income	1,505,854	1,792,584	1,990,606	2,180,779	2,205,987	2,114,723
Publications	244,766	195,846	203,093	152,116	160,624	147,769
NiGEM	376,630	422,451	418,538	348,240	411,550	405,733
Donations and others	67,702	100,408	96,128	78,875	63,424	99,772
Investment	169,884	145,275	116,902	128,271	111,589	116,729
Total income	2,364,836	2,656,564	2,825,267	2,888,282	2,953,174	2,884,726
% Research income	64.2	67.4	70.4	75.5	74.6	73.3
Total expenses	2,342,770	2,627,734	2,812,716	2,890,316	3,039,916	3,260,213
Balance	22,066	28,830	12,551	−2304	−86,742	−375,487

Source Data from financial statements supplied to the UK Government Charity Commission (2008–2013, Charity no: 306083), accessed 20 March 2016, http://apps.charitycommission.gov.uk/Showcharity/RegisterOfCharities/DocumentList.aspx?RegisteredCharityNumber=306083&SubsidiaryNumber=0&DocType=AccountList

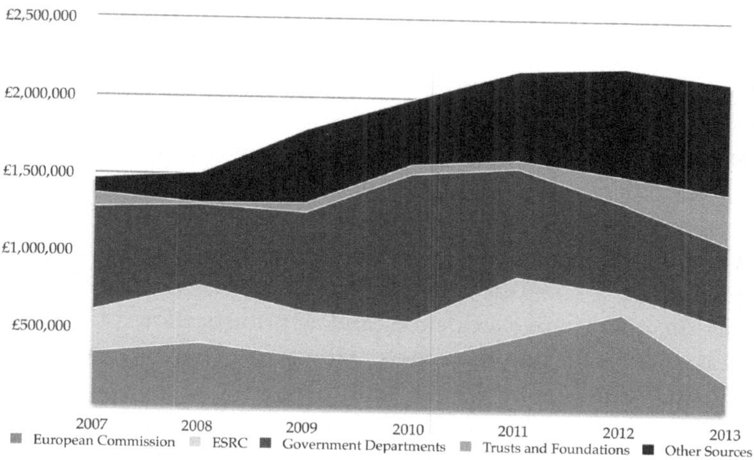

Fig. 5.1 NIESR research funding income per source (See source from Table 5.1)

permanent, senior) researchers. In view of the above, and as macroeconomics expertise became central to the public debate, one can venture NIESR had, following 2008, significant institutional threats and opportunities. NIESR's overall finances were unstable, while their reputation for rigour and impartiality and their research on macroeconomics—and later finance, migration and welfare—could make their work ever more influential, if publicly contested.

Style and Tropes

While recurrent tropes, styles and arguments are certainly present in NIESR's interventions, given their often-specialised character and the scope of authors published in NIER, these can be hard to detect—especially when compared with more overtly political organisations such as those covered in previous chapters. NIESR is, however, interesting as a case-study because, given its need to maintain a reputation for scientific rigour, it has to strike a balance between influence and technocratic distance. As will be shown later, NIESR's tropes, especially concerning the primacy of evidence, became ever more explicit as it faced a vague yet ubiquitous crisis of confidence in expertise, partly brought about by the perceived failure of mainstream economics to foresee the crisis.

Further, given NIESR's reliance on the interests and expertise of individual researchers—who frequently operate as if they were part of a 'university department,' each with different research priorities—continuities are at times hard to identify across policy areas. Nonetheless, these exist, conspicuous in a scientific ethos, a strive towards normative neutrality, a focus on quantitative evidence, and a prioritisation of certain policy areas and problems over others. However, given the array of topics NIESR intervenes upon, its high staff turnover, and its specialist profile, what follows can only be an overview of their familiar tropes and styles.

On account of NIESR's traditional publics—economists (both academic and 'in the wild'), civil servants, politicians, City experts, and financial journalists—its influence in the policy debate is often mediated through other elite actors. Although their forecasts and indicators enjoy considerable coverage, a significant part of NIESR's work is either behind a paywall (NIER) or expressed in a language that prevents engaging general audiences with ease. Its social media presence was, at least until 2011, minimal, its communications team small, and its public events, albeit policy-oriented, often of an academic bent. Nonetheless, even if this distance with lay audiences was partly abridged after Portes' directorship engaged more actively with non-economists, there are striking contrasts between NIESR and most advocacy-oriented think tanks. Its public interventions are perhaps more akin to those of independent and public or semi-public institutions producing macroeconomic data and research—frequently with much greater financial backing. Organisations producing comparable work include the IMF and the OECD—whose indicators and methodologies can act as a benchmark for NIESR's figures—specialist think tanks and research institutes (IFS, PSI), university departments and research centres (CFM, CEP), and even government bodies (ONS, OBR).

Given the above, it is worth pondering in what sense is NIESR comparable to the other case-studies covered by this book. Denham and Garnett (1998, 2004) count NIESR among early twentieth-century *avant la lettre* think tanks, which sought to be inspired by political neutrality and socio-scientific rigour.[18] More hedgehogs than foxes, NIESR could only awkwardly be put next to militant institutions such as 1970s new wave think tanks, or even those that, while partisan, make a point of being driven by evidence and seek to command a certain respectability with rival factions (e.g., IPPR, PX). A key reason for this is that NIESR does not chiefly produce clear-cut policy proposals, but rather policy-relevant research to inform the public debate. When its public interventions suggest a particular policy direction, they are often stated briefly, in the conclusions, and framed as derived from close data analysis. NIESR's recommendations often take the form of 'hedged' deductions, as opposed to blueprints in the manner of ASI's 'policy-engineers.' To give but one example, a 2012 NIER article proposed a more vigorous fiscal stimulus through the following:

> [I]t can be argued that fiscal policy choices have to be considered in the light of the monetary policy response function. When monetary policy is constrained by the zero lower bound on interest rates, the impact of fiscal policy (the fiscal multiplier) will be magnified compared to normal times. (Bagaria et al. 07/2012: f51)

One would also be hard-pressed to find in NIESR's public interventions normative arguments over what government should strive for, besides stressing the need for cogency between aims and methods—whether policies are sound and effective in their own terms—and an implicit preference for boosting growth, employment, and productivity rather than defending abstract principles such as 'market freedom.' Perhaps the above explains this reflection by an interviewee: 'we always think of ourselves as a research institute rather than a think tank [...] we just get called a think tank [...] but I don't know if we do or we don't fit in' (NIESR interview).

[18]This type of organisation finds parallels in the United States in the faith in positivist social sciences implicit in early twentieth century institutions such as the Council of Foreign Relations (CFR) and the aforementioned NBER (Medvetz 2012a).

Still, for the purposes of this book, NIESR provides an interesting point of comparison to more overtly political case-studies. Furthermore, economic indicators could be considered public interventions in their own right, given their effects and the epistemic choices they necessitate (Alonso and Starr 1987; Porter 1995; Eyal and Levy 2013). And although openly ideological think tanks produce indicators, these are rarely as grounded in academic economics as NIESR's, which strengthens their epistemic authority across audiences. For instance, it is unlikely that NEF's *Happy Planet Index* would be cited as evidence by the ASI—or that ASI's *Tax freedom day* would be by NEF—but both organisations have used NIESR data to substantiate their own claims.[19]

On account of this advantage, and of its committed defence of its impartiality, one could posit that, similarly to the IFS, NIESR occupies the role of 'umpire' of policy, or perhaps in Foucault's (1980) terms, of 'specific intellectuals' that intervene publicly on account of their specialised knowledge. In that respect, NIESR members are comparable to Pielke's (2007) 'science arbiters,' seeking to inform policy from a non-normative standpoint by virtue of the possession of expertise other actors do not possess.[20] Yet, as shall become clear, NIESR faced pressures, for instance with regard to funding, to become a more typical think tank with a more unified image and message.

To maintain a reputation for cognitive autonomy, NIESR needs to be perceived to be at a distance from its funders, a sizeable part of which are public bodies. Hence why its management has historically spurned most forms of core funding and patronage, which interviewees

[19]See NEF's (Reid 04/2013) and ASI's (Oliver 01/2012) in their respective chapters (III and IV).

[20]"The Science Arbiter seeks to stay removed from explicit considerations of policy and politics [...] but recognizes that decision-makers may have specific questions that require the judgment of experts, so unlike the Pure Scientist the Science Arbiter has direct interactions with decision-makers. [...] A key characteristic of the Science Arbiter is a focus on positive questions that can in principle be resolved through scientific inquiry. In principle, the Science Arbiter avoids normative questions and thus seeks to remain above the political fray" (Pielke 2007: 16).

believe could undermine their reputation as independent critics.[21] The Institute's founders themselves were adamant on its autonomy, as at the time of its inception they sought to meet the need for "someone outside government [...] in a position to challenge the analyses by the Treasury" (Denham and Garnett 1998: 65). One could list as a testament to NIESR's strive towards independence their many public interventions criticising official policy, even when not institutionally convenient. For instance, the Institute staunchly opposed Geoffrey Howe and Nigel Lawson's budgets in the 1980s, at the same time as the Thatcher government threatened cuts to SSRC funding—then, as the ESRC is now, one of NIESR's main sponsors (ibid.: 72). Years later, NIESR director Martin Weale lambasted the Labour Treasury for 'fudging' its own fiscal rules in *The Times* (2004). It could hardly be otherwise after 2008. Across the seven years of this study, NIESR disapproved of both Labour and Conservative spending plans and was critical of the European Commission (Portes 04/2012), all while continuing to receive funding from all these sources.

Such a position would be unsustainable without strong claims to specialised expertise, as the interests of the organisation should be seen solely as those deriving from the rigorous pursuit of evidence. Not casually, NIESR is well grounded in academic economics, and well placed to speak on behalf of the epistemic community of mainstream economic experts. To maintain this position, their judgement needs to be widely respected, which hinges on NIESR's perceived cognitive autonomy—and, in Medvetz's terms, a reliance on academic capital, which necessitates a distance from direct political pressures. Hence why, not unlike NEF and ASI in previous chapters, NIESR needs to define the limits of what should be considered sound economics—arguably in their case with a greater basis in academia. This effort requires boundary-work

[21]"Our independence is everything, our reputation for independence. So that's something that does worry us [...] if they begin to try and attack that. [...] We are deliberately not aligned with any political party or any political stance. We value [...] independence in our thought and [...] if you've got core funding it's probably increasingly more difficult to have it. But it makes fundraising easier if you have that [...] income on a regular basis" (NIESR interview).

over what economics is: defending the 'epistemic supremacy' of academic experts and determining which economic theories should be ignored. Indeed, many public interventions by NIESR explicitly delineate what the discipline's consensus is, where it exists, and what the best available evidence suggests (e.g., Portes 03/2013).

A further illustration of this strive towards socio-scientific legitimacy is that, on account of the importance for NIESR of being seen to produce sound macroeconomic research, most of its publications disclose the underlying assumptions that are built into their models. In such a way, NIESR showcases intellectual transparency by making explicit their past forecasting errors, while seeking to improve their predictive power. To that effect, NIESR has published research assessing the performance of its own forecasts (Kirby et al. 05/2014). This openness, however, implies an obverse risk: by its very nature, forecasting can be criticised for its past inaccuracies. Partly because of that potential weakness, NIESR's form of intellectual change is different from that of more advocacy-oriented think tanks and more grounded in the technical aspects of economics.

Furthermore, being NIESR, in Medvetzian terms, at the crossroads of academia and policy, differences between the orientation of staff are sometimes visible—between greater visibility or scholarly closure, between distance and engagement in the policy debate. One interviewee commented:

> It's fair to say there's [some] who are quite academic, who want to produce academic outputs. But certainly the director [...], most of the senior staff, and a lot of the junior staff [...] are motivated by having an impact on policy. It's not research that's for academics to read, it's research for policymakers to read and think about. (NIESR interview)

Moreover, following 2008 and the ensuing discredit of economic experts, intervening as 'ivory-tower' intellectuals became increasingly problematic. In that juncture, the Institute continued to strive towards neutrality, even as complaints about its purported partisanship grew, often linked to a faint but common branding of NIESR, especially among right-of-centre circles, as 'Keynesian.' In that precarious positioning, between being advisers and arbiters of the powers that be, NIESR straddled through a crisis of economics, politics, and technocracy.

'What the Data Says': The Politics of Cognitive Autonomy

Earlier in this book, while expanding on the *hysteresis* hypothesis, I ventured that, for a think tank seeking a reputation of non-partisanship and rigour, dispassionate warnings after the crisis' initial signs would likely be followed by assessments of its potential economic consequences. This discourse would presumably be continued by a condensation of the Institute's recommendations, judgements of the dominant policy agenda, and repeated attempts for its data to be considered and its advice to be heeded. This section will examine NIESR's public interventions between 2007 and 2013, with the aim of assessing to what degree they diverged from such expectations.

To the above, one should add that NIESR's GDP estimates and forecasts could be considered public interventions in their own right. This inclusion is based on Eyal and Levy's (2013) point that producing 'opinion' is a limited way of examining the influence of experts in the public debate. This is because it excludes techniques in the arsenal of experts (e.g., statistics, indicators, rankings) that are often more impactful than, for instance, editorials or blogs.[22] Additionally, NIESR's reliance on quantitative data for their public interventions can involve a certain distance between how they themselves and others interpret their work. As economic indicators often are perceived to 'speak for themselves,' the planning and effects of their release are more difficult to control to that of more 'textual' policy reports: they are commonly re-construed, cherry-picked, and challenged, and their implications reinterpreted.[23]

[22]"[I]ntervention cannot be efficacious without being equipped with all that makes expertise strong, and that opinion by itself lacks, namely, techniques, instruments, demonstrations, figures, charts, numbers" (Eyal and Levy, op. cit.: 228).

[23]"Once [our data] moves into the public domain then you have no control over how your message evolves. So it is very common for us to put pieces out and people to take not quite the extreme, but at times almost [...] diametrically opposed views on the basis of the same piece of work. But that happens with statistics. Statistics come out from the ONS and all sorts of different messages are spun from exactly the same number. We try and be very clear in what we're doing and how we're interpreting our work, but [...] if someone wants to use it in a certain way they will. You can put a formal piece out saying we would like to clarify that's not what it says or 'would you correct that,' but once it's out there, it's out there" (NIESR interview).

One could also say these numbers are 'performative,' playing a part in the reality they describe (Callon 2007). Moreover, by the very nature of forecasters' trade, numbers can showcase past errors, which can undermine their position in the eyes of critics. A few instances of internal reflections on the performance of their data and forecasts are scattered in what follows, although the various applications and re-appropriations of these numbers could inspire a research project of its own.

Before the crisis' onset, NIESR had been broadly supportive of Labour governments' economic record, yet critical of its low savings and of the UK's modest productivity growth (Barrell et al. 04/2005). NIESR did not outright predict a financial nor a fiscal crisis in the UK, although it issued warnings in NIER of growing inflationary pressures and argued there was room for fiscal tightening and greater savings (Kirby and Riley 01/2007; Barrell and Kirby 04/2007). In the global economy, NIESR forecasted overall growth with a modest slowdown in the United States due to weaknesses in its housing market. In April 2007, they claimed that '[t]here is some concern that [US] housing investment downturn may spread to other economies' (NIESR 04/2007: 7).

This level of alarm changed after Northern Rock's panic in September 2007, as did NIESR's policy focus. The first bank run in Britain since 1866 was bound to cast a light on the financial industry's vulnerabilities. As a result, some issues explored by NIESR in late 2007 included the tripartite character of British banking regulation—depending then on the BoE, the Treasury, and the Financial Services Authority (FSA)—the US slowdown, and increasing debt imbalances in the world economy (Weale 10/2007). Commenting on Northern Rock, the then Director said to the BBC (2007): 'there is a risk that financial services will contract particularly rapidly. The next couple of years are unlikely to be boring.'

Much of NIESR's public interventions in early 2008 claimed that, while the UK would be adversely affected by a more volatile financial environment, it was improbable it would fall into recession. The latter, however, was now a possibility. Those working on the NiGEM model were especially active at the time, studying what were expected to be worsening economic conditions, fuelled by a troubled credit market and rising oil prices. Different econometric conjectures were made on the possibility of a recession, which in hindsight proved optimistic. The January 2008 edition of NIER stated that:

> [a] severe US banking crisis would entail some combination of shocks. In our last Review [...] we considered a banking crisis scenario that involved a 4 point rise in global investment risk premia, a 2 point rise in consumer borrowing spreads, a rise in equity risk premia, a fall in US house prices of 6 per cent and a fall of about 3 per cent in European house prices. A shock of this magnitude would drive the US economy deep into recession, with the level of US output expected to be stagnant this year relative to 2007, and growth recovering to baseline rates only in 2010. Euro Area growth would fall below 1 per cent this year, and the UK economy would expand by less than 1⁄2 per cent. While we do not anticipate a major banking crisis of this magnitude, if perceptions of risk rise sharply and are sustained, this could push the global economy into a deep recession. (NIESR 01/2008: 14)

In April that year, Barrell and Hurst (04/2008) gauged again the potential effects of a crisis in the US banking system on the world economy, hinting at growing concerns over the state of the global financial industry. Nevertheless, there was no consensus a crisis was inevitable, as Weale claimed: "the economic outlook is surprisingly rosy, reflecting the fact that banking is not the whole of the economy" (Weale 04/2008: 8). While predicted UK growth for 2008 was downgraded to 1.5% in July (NIESR 07/2008)—compared to roughly 4% across 2007 forecasts—a recession was deemed unlikely "provided there is not another Northern Rock" (*Guardian* 2008).

There was, of course, another Northern Rock. Following US policymakers' refusal to bailout Lehmann Brothers, the last quarter of 2008 saw an acute contraction of the global and British economies. As many have noted, this policy decision and the downturn that ensued hurt the reputation of macroeconomists and forecasters, including NIESR—if, arguably, predicting the precise occurrence of an economic crisis is nigh impossible. In their words,

> the past year has obviously been a difficult time for the economy and the reputations of economic forecasters such as the National Institute have suffered from their failure to forecast the recession. (NIESR 08/2009: 2)

While acknowledging past errors, NIESR members pondered whether they could have done better, claiming: "[a] complaint that forecasts are 'wrong' of course is completely beside the point. The more relevant issue is whether better forecasts could have been produced using methods different from those we employ" (NIESR 08/2009: 3). In more technical terms, the inability of most experts to foresee the crisis, NIESR researchers believed, was partly due to the reliance of most models on 'Value-at-Risk' historical data, which provides few guides for unfolding events (NIESR 08/2008: 2). Members of the Institute were quick, nevertheless, to indicate that the lack of a peremptory policy response only made matters worse:

> We believe [..] this recession was inherently avoidable even in September 2008, and [...] rapid recapitalisation or even nationalisation of failing financial institutions should have come a few weeks earlier than it did. (Barrell 10/2008)

Locating the roots of the crisis in a failure of regulatory oversight, Ray Barrell, then Senior Research Fellow, claimed: "[faults] do not lie primarily with either fiscal designs or monetary structures, but with the failure to address the need for macroprudential regulation" (Barrell 10/2008: 2). Forecasts and GDP figures had to be revised. In January 2009, NIESR (01/2009) estimated the British economy shrank by 1.5% between September and December 2008, and in April, NIESR (04/2009) claimed that the prospects for 2009 were of a 4.3% contraction, followed by 0.9% growth in 2010. An accompanying decrease in tax revenue would strain the UK's fiscal situation, aggravated by similar developments in the Eurozone and the United States, constraining liquidity and demand across the board.

In such a context, and as one of the oldest macroeconomic research centres in Britain, NIESR became, almost by default, prominent in the public conversation. The role of NIESR as arbiters and forecasters made their work particularly relevant, even while beset by a growing mistrust of economic expertise. Precisely when forecasters such as NIESR were most vulnerable to criticism, their public profile rose and their potential influence grew. An index of this paradox is a noticeable increase in their appearances on broadcast media, reported in their own annual accounts (Table 5.2)—curiously accompanied by a relative drop after 2010 of more narrowly academic outputs.

Table 5.2 NIESR communications output

Type of output	2007–2008	2008–2009	2009–2010	2010–2011	2011–2012	2012–2013
Research reports, articles, book chapters	112	163	179	119	95	76
Conference and seminar presentations	41	27	93	69	56	53
Appearances on broadcast media	32	66	124	279	238	237

Source Data from Table 5.1

Concomitantly, the 2008 crisis generated shifts in NIESR's policy focus, partly due to new funding restrictions in some of the areas spurring previous research—sometimes dramatically so, requiring projects to be abandoned.[24] All the while, the demand for NiGEM rose, due both to the need to monitor a fragile recovery and to requests by financial regulatory bodies—such as the then Financial Services Authority (FSA; disbanded in 2013) (NIESR 08/2009: 4)—to measure the effects of potential regulatory changes in the financial system.[25] NIESR's 2008 annual report summarises these tendencies:

> The immediate effect [...] has been to increase interest in our macroeconomic work. Media activities have increased and we have also been approached to undertake research in which there would have been little interest a few months ago. (NIESR 08/2008: 1)

As the import of macroeconomics grew, and in order to inform a growing public interest in associated issues, NIESR needed to defend its record and the reputation of academic economics more generally. The Institute became particularly important as a judge of when the British economy would resume growing, and in this respect, their prognoses often complemented or were contrasted in the media with those of the IMF (BBC 2009a). The level of attention of their press releases remained considerable and in October 2009 NIESR estimated the British economy would contract by 4.4% that year (Barrell et al. 10/2009). However, even if by January 2010 NIESR had told the

---

[24]"From the perspective of those that work on macroeconomics at the Institute [...] particularly around NiGEM [...] the crisis really did change what we were working on [...] Running up to the crisis we were about to undertake a major piece of work just looking at macro-level consumption functions across Europe. We'd done all the data work [...] just to get things into the situation we wanted them to be. That was scrapped entirely because we weren't going to get [...] funding for it [....] so the cost of all the data work was gone. We've never gone back to it" (NIESR interview).

[25]"[After the crisis] we re-focused entirely. We started having calls from various institutions [...] The Financial Services Authority as they then existed [...] came to us and [said] 'we [...] provide you with funding, can you try and prepare a banking sector model for the UK into NIGEM that we can use?' [...] they wanted to [...] gauge what impact regulatory changes to banking sector would have on the real economy" (NIESR interview).

Financial Times (2010a) the UK was out of recession—meaning the last quarter of 2009 saw 0.3% growth—it was still the case that the recovery itself (i.e., returning to the GDP per capita levels of before the crisis) would take until 2014 (BBC 2009b). For that reason, far-ranging measures were needed to boost growth and confront a debilitated fiscal position.

Throughout 2009, NIESR explored different options to offset the fiscal imbalance that followed the recapitalisation of banks and the tax revenue shortfall, concentrating especially on three avenues. First—and partly due to demographic trends—NIESR researchers advised to consider extending working lives by increasing the pension age (Barrell et al. 2009). Second, NIESR enquired into tax hikes to tackle a sharp rise in the public deficit—being critical of a temporary VAT reduction under Gordon Brown (Barrell and Weale 03/2009). Third, NIESR examined different models of spending cuts and their effects on demand (Barrell 10/2009a). However, none of these measures in isolation would suffice. In mid-2009, Research Fellow Simon Kirby put the matter to the media:

> As a country we are faced with a decision [...] There must be a combination of spending cuts, higher taxes and longer working lives as 'each one, in isolation, is too extreme an option'. (Kirby in *Financial Times* 2009)

Concerning spending cuts—ever more politically central given the Conservative's austerity platform in the 2010 elections—the Institute had argued that there was space before the crisis to consolidate fiscal policy and to increase savings (Weale 10/2008, 2010), especially on account of the need for intergenerational fairness (Barrell and Weale 09/2009). Nonetheless, in October 2009 NIESR criticised then Shadow Chancellor George Osborne both for errors in his estimated budget cuts and his misuse of NIESR data (*Guardian* 2009). More broadly, although NIESR researchers were supportive of the need to tackle the deficit, the pace of adjustment would elicit important discrepancies.

In parallel, and in consideration of regulatory failures leading to the crisis, NIESR explored policy options to bolster a damaged financial system. This branch of their work was ever more important, as a likely rise in risk premia[26] would scar growth prospects (Barrell 10/2009b). Across its publications, NIESR argued for increasing banks' capital requirement (Barrell et al. 2010), a measure that, while slightly detrimental for growth, substantially reduced the possibility of another crisis (Barrell and Davis 04/2011). Such research has informed the recommendations of Basel's Bank for International Settlements, the FSA, and the ICB (NIESR 03/2011).

In April 2010, NIESR governor and former MPC member Alan Budd, who helped set up the OBR,[27] published in NIER a paper criticising Labour government spending which, uncoupled with tax rises, lead to a 3% deficit which surged to around 11% after the crisis (Budd 04/2010; see also *Financial Times* 2010b). The government had failed to save, respect fiscal rules, and prepare itself for a possible downturn. A month later, the election of a Conservative-led coalition settled the official policy response to the crisis: deficit reduction through public-sector cuts. However, while NIESR was initially supportive of some fiscal tightening—with the proviso of not offsetting the recovery—they warned against the Coalition's proposed cuts from early on (*Financial Times* 2010c). In June 2010, NIESR claimed the British economy had returned to long-term growth trends (around 0.6%), but that prospects remained uncertain given the Eurozone's instability and as the effects of austerity were yet to be felt (*Financial Times* 2010d).

In parallel, two crucial institutional transformations were under course. The first concerned finances. Changes in the funding climate had effects on NIESR's staff and resources, accentuated by the Institute's relative dearth of core donations. This sensitivity to pressures from the funding environment, coupled with constraints derived from dwindling government spending on research—which had prompted diversification efforts for some years (NIESR 08/2007: 3)—meant NIESR went through a period of sustained losses, having to rely on its savings.

---

[26]A risk premium is the "additional expected return that an investor requires over the return on a risk-free asset" (NIESR 11/2009: 19).

[27]The Office for Budget Responsibility (OBR) was established in May 2010 as an advisory public body tasked with producing independent assessments of the state of British public finances.

As such, and in consideration of austerity, the Institute's traditional dependence on public sources declined as it sought to attract more charitable and private sponsors. One interviewee noted:

> We made a decision [...] that we would move away from our heavy reliance on government department funding [because] we saw that there was going to be constraints in their research budgets [...] Also, there was a perception that government departments were asking for more and more for the funding that they offered. And so we did make a decision to [...] diversify our funding base, to move away to charities, foundations, trusts. (NIESR interview)

This shift of strategy implied that, as different types of funding are linked to an organisation's position—e.g., ESRC grants depend on academic prestige, trust funding is tied to the advancement of charitable aims—NIESR required some changes in its image. In what could be interpreted as a case of isomorphism (DiMaggio and Powell 1991) NIESR moved closer to other think tanks in the British landscape, both in institutional terms, concerning their funding sources, and in their behaviour, towards more cohesive and active public engagement. One interviewee said:

> One the things that we're doing is [...] ensuring that our brand and our identity are stronger to the outside world. [S]ome of that is around impact, some of that's around funding so that we can increase the funding that we get for research projects. But also we're constantly aiming to get support that's not tied to research funding, so corporate donors [...]. We do have some corporate donations, but we don't have as much as we would like. (NIESR interview)

A second crucial institutional transformation, linked to the above, concerns a change of directorship. In August 2010, Martin Weale joined the BoE's MPC, and Ray Barrell became the interim director until Jonathan Portes' arrival in February 2011. With a background as a civil servant under Labour and Conservative governments—he was Cabinet Office Head of Macroeconomic Analysis in September 2008—Portes brought a different type of leadership to the organisation, concerned with playing a pivotal and concerted part in the policy debate:

> Martin's directorship [...] was very much 'everyone can do whatever research they want,' so he didn't have any real opinion [...] on whether somebody should be doing research in any kind of different areas [...] Jonathan's view, and I think the senior staff's view, [...] now is that actually we need to act more as an organisation rather than a series of individuals. (NIESR interview)

Portes—himself an active public figure—spearheaded NIESR's push for prominence through his many public interventions, both aimed at expert and general audiences. In the process, he often embodied the proverbial Foucauldian specific intellectual, speaking truth to power from the basis of specialised knowledge. This is the case even if (perhaps because) he has a hybrid profile compared to most NIESR staff, with comparatively more experience in the civil service than in academia. In time, Portes' visibility earned him and NIESR a reputation for being vigorous opponents of austerity, and the Institute became the 2011 *Prospect* 'think tank of the year.'[28]

The arrival of Portes also had an effect on the organisation's self-understanding. As one interviewee put it, from resembling a 'university department' where individual researchers are relatively free to decide how to spend their time, "now there's more of a sense that we should be a kind of corporate entity" (NIESR interview). That push for further integration implied the organisation had to coordinate its public interventions across research teams, arguably moving NIESR closer to the definition of think tanks I expanded on the introduction to this book: institutions 'on behalf of which' individuals intervene on public policy matters.

Greater efforts in coordination were accompanied with a push, in Medvetz's (2012b) terms, to become more of a 'boundary organisation' more firmly positioned in the fields of media, business, and politics. This shift is discernible in NIESR's quest to increase its media presence, add corporate donors, and in its numerous public appeals to policymakers. An interviewee compared the two directorships: "Martin was very

[28]Accessed 17 October 2015, http://www.niesr.ac.uk/sites/default/files/publications/131011_171942.pdf.

much academic [...] about publishing academic findings to influence policy. Jonathan is much more about engaging with policymakers using research" (NIESR interview).

In terms of the format of their public interventions, from 2011 onwards NIESR saw a rise in op-ed production and use of social media, and or in Schmidt's (2008) terms, of 'communicative discourse.' The training of NIESR researchers on how to use Twitter and the opening of an institutional blog are further examples of such a push for greater visibility and influence.[29] Starting in January 2012, NIESR's blog has published on their research and on politically salient issues, a substantial proportion written by Portes himself. In frequent conversation with other economists, both allies (Simon Wren-Lewis, Diane Coyle) and critics (Chris Giles, Andrew Lilico), NIESR's blog has become a focal point for the dissemination of 'expert opinion' and for linking otherwise separate parts of the organisation.

Many of Portes' public interventions criticised the Coalition's policies, which he deemed to be self-defeating. This applies both to government departments and to British macroeconomic governance, sometimes employing rhetorical devices such as irony rather than purely technical language—e.g., "DWP analysis shows mandatory work activity is largely ineffective: Government is therefore extending it" (Portes 06/2012a). One common theme in NIESR's blog is an insistence on measuring the effect of specific policies on growth, employment, and productivity while criticising the strict adherence to what is considered a fallacious narrative, dubbing the idea that government spending cuts bolster credibility to be 'faith-based economics' (Portes 11/2012). Arguing that, in recessions, debt-to-GDP ratios worsen after overzealous debt consolidation, NIESR claimed austerity "[e]ven on its own terms [...] is making matters worse" (Holland and Portes 10/2012: f4). Beyond the UK, Portes (04/2012) stated that, by impacting overall demand, austerity across Europe would hurt the growth prospects for all.

---

[29]"We re-vamped the website completely, introduced a blog post, went into social media, trained all our staff in social media and Twitter and we've aimed to have much more of an impact, and we have had an impact. So our profile has really risen in the last few years" (NIESR interview).

In a nutshell, NIESR argued against expansionary fiscal contraction—the theory that public spending reductions would not hinder economic growth and might even boost it—which they regarded as a minority view among economists but which enjoyed unjustified political leverage, attributed to what they considered a discredited paper by Alesina and Ardagna (2009). Portes (10/2012) said instead that the economic consensus was that austerity, especially after crises, would reduce demand by weakening fiscal multipliers. Furthermore, given the low cost of servicing UK government debt, official policy should instead favour expanding public investment.

In early 2012, after months of meagre growth, NIESR claimed Osborne's plans undermined the recovery and a change of course was needed (Portes 03/2012). In January 2012, NIESR predicted that, having grown below 1% in 2011, the UK would undergo a 0.1% contraction in 2012 (NIESR 01/2012: f3). Around that time, NIESR also published a series of graphs representing the monthly trajectory of recoveries after various economic crises in British history (NIESR 01/2012). They showed that, after a promising start, after austerity growth began to linger when compared to earlier recessions. These graphs received ample coverage and became a poignant illustration of NIESR's arguments and the reasons behind their opposition to the government's economic agenda.

In July 2012, NIESR (07/2012) predicted the economy would contract by 0.5% that year and grow by 1.3% in 2013. Towards the beginning of 2013, the economy was expected to grow again, albeit it was uncertain for how much. In February, NIESR (02/2013) claimed that, although 2012's GDP figures had been mostly flat, 2013 would see 0.7% growth. In August, NIESR revised its forecast back to 1.3% (NIESR 08/2013). Estimates proved to be slightly pessimistic, which according to interviewees might be related to errors in the underlying data.[30] Another intervening factor mentioned by interviewees is that,

[30]"We [...] became a bit too pessimistic in the second half of 2012 into 2013 because of real world events. In particular, our 2013 forecasts were low because of data [...] that were subsequently revised up quite significantly. So that led us down that path [...] I find that quite interesting because people say you were just continuously pessimistic and gloomy and it's like well, if you look at our 2012 forecasts, we weren't that far off where the economy got to" (NIESR interview).

although NIESR forecasts tend to perform well, researchers find little time to assess how they could be improved to the degree better-funded organisations can (IMF, OECD, OBR).[31]

Overall, the return to growth in 2013, Portes argued, was due to the effects of fiscal consolidation fading out and the fact that, even if the rhetoric of austerity was still present, the pace at which the deficit was being reduced was slowed (Portes 10/2013). However, although positive macroeconomic figures had returned, NIESR researchers claimed much terrain had been lost by pursuing rapid government-spending contraction. Ultimately, since 2007 "the performance of the UK economy has been poor" (Riley and Young 05/2014: r1).

Among the controversies surrounding austerity on which NIESR intervened, one of the most emblematic concerned the interest rate on government bonds, or gilt yields, and how their fluctuations should be interpreted. Portes declared that the low cost of borrowing, besides being an incentive for, precisely, borrowing, was caused by modest growth expectations rather than confidence that contractual commitments would be met, given that Britain pays its debts in its own currency (*New Statesman*, 2011). The UK lower interest rate, rather than being a sign of credibility, was symptomatic of economic weakness. Portes declared to the Treasury Select Committee in late 2012:

> [Gilt yields are] not about credibility [but] about what you think it's going to happen to the economy. Economic weakness leads to low long-term interest rates [...] this is really quite basic macroeconomics [...] All I'm pointing out is 'look, this is what basic economic theory says ought to happen, this is what empirical evidence says should happen, they coincide, it doesn't correspond with what the government is saying'. (ParliamentLive 2012)

---

[31]"So this year and last year [2014-2015] we've just tried to publish a very short piece that just gives our forecast errors [...] We performed better than some recent benchmark forecasts, started comparing ourselves to a few other institutions [...]. So we've not done [forecast error analyses] at the level of detail the OBR, the IMF, or the OECD have, which is fine because they've got the resources to do it [...] It is a resource question. It would be nice for us to do that, but no one is going to provide the funding for us to [...]. We may be able to carve out the time gradually over the years to look at that, but that's a luxury that I'm envious of" (NIESR interview).

Portes did not only spar with policymakers. Across our timespan, he debated through the media and social media journalists, think-tankers, politicians and commentators, particularly those supporting austerity or for restricting immigration—including Mark Littlewood from the IEA (*Prospect* 2013), David Goodhart from Demos (London Review of Books 2013), Andrew Lilico from Policy Exchange (ConservativeHome 2013), the conservative historian Niall Ferguson (*The Spectator* 2015a), Conservative MP Jesse Norman (Norman 2012), and Mike Fabricant, then Conservative Vice Chairman (Portes 01/2013). In these exchanges, Portes frequently presented himself as a scrupulous economist attempting to debunk those judged to be misconstruing data. In that exercise, he sought to maintain a reputation for cognitive autonomy for himself and NIESR, focusing on evidence and steering clear from direct normative advice. His efforts sometimes elicited scolding from opponents, to the point that NIESR was denounced as falsely centrist by Conservative MEP Daniel Hannan (Pieria 2013). However, in that case, as in others, Portes made their opponents retract or correct their assertions through the Press Complaints Commission.[32]

Throughout Portes' directorship, NIESR continued to work in areas impacting productivity, finding that Britain's had been lagging (Goodridge et al. 05/2013; see also NIER November 2013 issue). Given that productivity is considered to be 'the main, if not the only, driver of real wages and overall prosperity' (NIESR 05/2014) this was an issue of great relevance. As some of the main factors impacting productivity include education, training, migration, labour policy, investment, and innovation, these areas were often the focus of NIESR's research. For example, given Britain's demographic pressures, lifelong learning was considered a central theme for future economic performance (Dorsett et al. 08/2010). Hence NIESR's involvement in the ESRC Centre LLAKES.

Throughout the period this book considers, NIESR also intervened on policy-relevant issues as they arose, which generated new programmes, brought in staff and funding opportunities. For instance, in

---

[32]See (accessed 13 October 2015) http://www.pcc.org.uk/advanced_search.html?keywords=jonathan+portes&page=1&num=10&publication=x&decision=x&image.x=0&image.y=0.

2011 Angus Armstrong, former Treasury Head of Macroeconomic Analysis, joined NIESR as Director of Macroeconomic Analysis. One of his first publications was an assessment of the ICB recommendations on financial reform (Armstrong 10/2011). Towards 2013, Armstrong helped produce research on the economic consequences of an independence vote in the run-up to the 2014 Scottish referendum (see also NIER February 2014 issue). The above was accompanied by ever greater efforts to produce rigorous work that could be easily communicated to wider audiences through new forms of public engagement. For instance, in September 2013 NIESR produced its first YouTube video on currency options for an independent Scotland, seeking, all the while, to maintain a reputation for non-partisanship with regards to what was a very contentious referendum.[33] Similar undertakings would continue beyond 2013 as Britain's membership to the EU became more of a concern, for instance through Armstrong's and Portes' participation in the ESRC Centre 'UK in a Changing Europe.' Furthermore, as NIESR had been among the first to explore the economic repercussions of a hypothetical UK withdrawal from the EU (Pain and Young 2004), it was in a privileged position to produce further public interventions on the matter as the issue gained traction (Portes 11/2013).

Another area in which NIESR became increasingly prominent is immigration. NIESR researchers argued there is a positive correlation between productivity, growth, and high-skilled migration (Hierländer et al. 2010), which set NIESR against a rising political tide seeking to restrict net migration figures, which included the Government. Indeed, the very first NIESR blog post took issue with Migration Watch figures (Portes 01/2012a). Throughout these years, and in line with the Institute's positioning as independent experts, its work on migration became ever more important. On this topic, NIESR sought to be regarded as politically neutral, providing the same type of evidence-driven recommendations that characterised their work on macroeconomics. Furthermore, like the financial crisis, the growing political import of immigration offered new avenues of research funding. In 2013 the Foreign Office commissioned

[33]Accessed 1 November 2015, https://www.youtube.com/watch?v=mBC0mLFz91o.

NIESR for a report on the lifting of work restrictions to Bulgarian and Romanian citizens after January 1, 2014, which concluded that, notwithstanding the lack of reliable data, economic migrants from those countries were unlikely to head primarily to the UK (Rolfe et al. 04/2013). Through such public interventions, NIESR began to partake in an increasingly bitter public debate while still seeking to be seen as impartial (e.g., *Guardian* 2013)—and, much like in economics, this mission became ever more difficult as years went by.

Overall, despite its history and reputation as independent experts and arbiters, NIESR's actual policy impact is hard to trace. One possible exception is their work on caste discrimination leading up to an amendment to the 2010 Equality Act (Metcalfe and Rolfe 12/2010). However, in what concerns fiscal policy and immigration, the Coalition's measures were, not infrequently, exactly the opposite of what NIESR recommended. Furthermore, although the Institute continued to be contracted by government departments, this type of funding became less central as other sources were increasingly pursued—partly, as interviewees claimed, since the competition for such research commissions had increased.[34] Beyond the timeframe of this book, due to what is reported by a NIESR governor to be internal conflicts arising from funding shortages (Medium 2015), Portes ceased to be NIESR's director, much to the glee of some of his detractors (*The Spectator* 2015b).

Umpires in Unruly Arenas

I wrote this chapter minding Eyal and Levy's (2013) observation that the public interventions of experts should be understood more broadly than within the bounds of 'opinion.' Much like economists can partake in the policy debate in more ways than through expert commentary *à la* Krugman, NIESR has more tools at its disposal than words alone. The attention its GDP estimates and forecasts enjoyed in the wake of the

[34]"[On government department research] the competition has increased because there's a lot of different organisations now bidding for that kind of work, which there weren't in the past" (NIESR interview).

crisis is a case in point. Yet, after 2011, NIESR started producing more and more opinion pieces, whether in traditional print and broadcast media, social media, or in parliamentary consultations. In that sense, although the *hysteresis* hypothesis from Chapter 2 proved a relatively accurate guide for NIESR's unfolding view of events—highlighting the centrality of evidence and seeking to provide it—much like with NEF and the ASI, it said little on how the formats and modes of engagement of their public interventions shifted.

On this point, a noticeable growth in the coverage of NIESR's econometric indicators, and the increased production of media commentary that ensued, are likely linked to a discordance between what NIESR staff claimed economic evidence suggested and the actual employment of economic data by the media and policymakers. Throughout 2009–2011, their policy recommendations, however carefully based on rigorous research, went unheeded, while what they deemed a 'faith-based economics' became prevalent. In such a context, intervening as aloof experts insisting on the primacy of evidence became hardly politically innocuous. The line separating 'experts' and 'specific intellectuals'—those who produce evidence to inform policy and those who criticise political power based on evidence—became blurry. Hence, NIESR's public engagements tended to elicit accusations of partiality from its opponents, in a milieu in which a reputation for neutrality became ever harder to maintain. In other words, the strategy of *de-politicising* policy that Diane Stone referred to in the preface to this book became increasingly ineffective. For good or ill, macroeconomic evidence came to be seen as inherently political, and therefore biased.

A recurrent trope in the sociology of expertise has been the ubiquitous rise of economists to positions of power (Markoff and Montecinos 1993). However, at least since Thatcher's dismissal of the advice of the 364 economists who wrote to *The Times* in 1981 to criticise her policies (Norpoth 1991), the relationship between academic economics and politics has been more fraught than frequently assumed. Weingart (1999) spoke of a 'paradox' at the interface of science and politics: at the same moment scientific advice is most required politically, it is most prone to be accused of politicisation. Come 2008, as shown by NIESR's case, the advice of economists—or at least of the economic consensus as construed by those with the most academic gravitas—seems to have been less influential in policy than it would be commonly thought.

Concerning the uses of evidence for policymakers, Boswell (2009) has argued that the role of research often tends to be, rather than instrumental, symbolic: substantiating a policy preference and legitimising its proponents. If Boswell is correct, NIESR's research was bound to be overlooked when unhelpful to the official agenda, as a scientific ethos implies a necessary distance from political expediency. In her words, an "important discrepancy between the systems of politics and science" (ibid.: 97) is then likely to ensue. Hence perhaps NIESR's move from prioritising more academic outputs and more 'coordinative' discourse to a more spirited engagement with social media and media outlets. However, in the context of a crisis of expert authority, *ad hoc* argumentation was less susceptible to be effectively debunked: with fewer 'epistemic authorities' recognised by all, fewer can arbitrate.

At the same time—and leaving aside the question over the scientific status and neutrality of macroeconomics—a divergence between the demands of politics and research meant NIESR's more engaged criticisms could only be considered politically partial by supporters of austerity, especially when done through mainstream and social media. In Bourdieusian language, this mismatch between fields might explain why NIESR was often depicted as Keynesian by its detractors—that being, foremost, a political judgment. Curiously, in a blog post entitled *Fiscal Policy: what does 'Keynesian' mean?*, Portes (01/2012b) argued Keynes should first be considered a scientific rather than a political inspiration: "[y]ou might as well have asked a physicist if he was a 'Newtonian.'"

The tension between being positioned as neutral observers while attempting to influence policy in a contentious political context also has its manifestation in internal debates over the public role of macroeconomists, which led to much pondering over the public status of their own profession (e.g., Portes 06/2012b). As specialised actors seeking to influence policy based on economic consensus and evidence (inasmuch as they exist), NIESR would try to maintain a reputation for producing specialised, empirical, and non-normative arbitration, regardless of how politically charged the issues under examination were. Nonetheless, in fractious public arenas and after a crisis of experts' credibility, that was a difficult status to sustain, and one opponents were eager to challenge.

In November 2012, as Portes testified in Parliament, he was accused of pulling NIESR to the left—incidentally, by Conservative MP Jesse Norman, a member of NIESR's Council of Management and Policy Exchange collaborator. Portes responded:

> [What I say publicly does] reflect the institutional view [of NIESR] we discuss these issues, we come to a consensus, and when I talk publicly [...] or write articles I write them on my capacity as director of the institution. Do they have some political significance or import? That's not for me to say but I wouldn't be surprised. [...] Macroeconomic policy wasn't so much five years ago but it is now, [...] rightly, a subject of public and political debate. If we say what we think the economics says [...], that may have political implications but that's not why we say it, we say it because it's our job. (ParliamentLive 2012)

When questioned, Portes claims the political significance of their work is derived from the weight of evidence rather than from any political bias. As that view cannot but be uncomfortable for those believing austerity is an objective necessity, they sought to undermine it. This conflict conjures what might be the crux of the precarious place of 'scientific arbiters' in politics: notwithstanding one's opinion of the scientific pretensions of economics, for a discipline 'obsessed' with governance (Bockman and Eyal 2002: 322), what is to be done when expert advice goes unheeded but to criticise and engage more vigorously in the policy debate? And how to avoid being seen as partial in the process?

I do not wish to solve that quandary here, but merely to note that NIESR's attempts to be seen as unbiased were bound to be questioned in an ever more splenetic public conversation, and one ever more mistrustful of expert authority. As a consequence, here as with NEF and the ASI, changes in the format of their public interventions were connected to efforts to reach and convince larger publics. In NIESR's case, this was further compounded by the fact that the reputation they sought for themselves was of political agnosticism and scientific rigour *while* taking a position on policy. On that regard, it is interesting to note that standing next to NEF's James Meadway in his Occupy LSX address (see Meadway 11/2011, in Chapter 3) is Jonathan Portes.

This tension between neutrality and political engagement is, however, not a new problem. An interviewee noted:

> If you look way back to when we were formed 77 years ago, there's very much a debate then about what NIESR should be about. Should we be trying to influence policy directly, indirectly, were we academic, were we policy people? You know it's interesting that is still continuing actually, there's probably never been a period in NIESR's history where we haven't had that debate. (NIESR interview)

Throughout 2007–2013, I have also shown how NIESR faced isomorphic pressures, pushed by a funding environment characterised by austerity. Nevertheless, important differences remained with advocacy-oriented think tanks. NIESR can still—at least to a greater degree than most other politically oriented policy institutes—be listened to across the spectrum, yet it can do little to control how its public interventions are interpreted. And, albeit the Institute remained influential and its work continued to be demanded, its position as government advisors—at least in what concerns macroeconomic management—was challenged by actors who were better connected to political parties. Arguably, at least since the technocratic ideal of the post-war period waned in the 1970s, other more overtly political think tanks have menaced NIESR's place in influencing policy, which might help explain its relative shift from 'coordinative' to 'communicative' discourse (Schmidt 2008; Ladi 2011).[35]

Overall, the distance between NIESR's recommendations and the government's policy response speaks of a growing divergence between what is considered dominant in the fields of knowledge production and politics. As part of the tensions NIESR confronts, being at the boundaries of academic economics and public policy, one should note that

[35]Denham and Garnett argue a rift between NIESR and government had already happened in the Thatcher years: "[a] new low point […] was reached in the early 1980s, when […] NIESR, which generally favoured growth, confronted a government which had committed itself to a policy which deepened an already serious recession" (Denham and Garnett 1998: 78).

what is dominant in one sphere of society can be peripheral another. Ideas that are marginal among academics can hold sway among politicians, and vice versa. I shall expand on the conduits between politics and expertise in the following chapter, as I describe a think tank that is tightly linked to political elites.

References

Alesina, A., & Ardagna, S. (2009). *Large changes in fiscal policy: Taxes versus spending* (NBER Working Papers, 15438). Accessed 10 April 2015. http://www.nber.org/papers/w15438.pdf.

Alonso, W., & Starr, P. (Eds.). (1987). *The politics of numbers*. New York: Sage.

Barrell, R., Davis, P., Karim, D., & Liadze, I. (2010). Bank regulation, property prices and early warning systems for banking crises in OECD countries. *Journal of Banking & Finance, 34*, 2255–2264.

Barrell, R., Hurst, I., & Kirby, S. (2009). The macroeconomic implications of pension reform. In D. France (Ed.), *Pension reform, fiscal policy and economic performance*. Rome: Bank of Italy.

BBC. (2007). *Northern Rock: Expert views*. Accessed 20 March 2015. http://news.bbc.co.uk/1/hi/business/6999246.stm.

BBC. (2009a). *UK recession 'worst since 1930s'*. Accessed 28 October 2015. http://news.bbc.co.uk/1/hi/business/8034879.stm.

BBC. (2009b). *UK recovery 'to take five years'*. Accessed 28 October 2015. http://news.bbc.co.uk/1/hi/business/8162217.stm.

Bockman, J., & Eyal, G. (2002). Eastern Europe as a laboratory for economic knowledge: The transnational roots of neoliberalism. *American Journal of Sociology, 108*(2), 310–352.

Boswell, C. (2009). *The political uses of expert knowledge: Immigration policy and social research*. Cambridge: Cambridge University Press.

Budd, A. (1999). Learning from the wise people. *The Manchester School Supplement, 1463–6786*, 36–48.

Callon, M. (2007). What does it means to say that economics is performative? In D. MacKenzie, F. Muniesa, & L. Siu (Eds.), *Do economists make markets? On the performativity of economics* (pp. 311–356). Princeton, NJ: Princeton University Press.

Coats, A. W. (Ed.). (2000). *The development of economics in Western Europe since 1945*. London: Routledge.

ConservativeHome. (2013). *Andrew Lilico: What Jonathan Portes gets wrong and why*. Accessed 15 October 2015. http://www.conservativehome.com/platform/2013/01/andrew-lilico-what-jonathan-portes-gets-wrong-and-why.html.

Daily Telegraph. (2009). Sir Bryan Hopkin. Accessed 20 March 2016. http://www.telegraph.co.uk/news/obituaries/politics-obituaries/6638037/Sir-Bryan-Hopkin.html.

Denham, A., & Garnett, M. (1998). *British think tanks and the climate of opinion*. London: UCL Press.

Denham, A., & Garnett, M. (2004). A 'hollowed out' tradition? British think tanks in the twenty-first century. In A. Denham & D. Stone (Eds.), *Think tank traditions* (pp. 232–246). Manchester: Manchester University Press.

DiMaggio, P., & Powell, W. (Eds.). (1991). *The new institutionalism in organizational analysis*. Chicago: University of Chicago Press.

Evans, R. (1999). *Macroeconomic forecasting: A sociological appraisal*. London: Routledge.

Eyal, G., & Levy, M. (2013). Economic indicators as public interventions. In T. Mata & S. Medema (Eds.), *The economist as public intellectual* (pp. 220–253). London: Duke University Press.

Financial Times. (2009). Think tank says cuts will not pay debt. Accessed 15 October 2015. http://www.ft.com/cms/s/0/39c54894-7655-11de-9e59-00144feabdc0.html.

Financial Times. (2010a). UK out of recession, says think tank. Accessed 23 June 2015. http://www.ft.com/cms/s/0/271f57e6-0054-11df-8626-00144feabdc0.html.

Financial Times. (2010b). Labour 'errors' attacked by former official. Accessed 15 October 2015. http://www.ft.com/cms/s/0/d86bed74-4f38-11df-b8f4-00144feab49a.html.

Financial Times. (2010c). Spending cuts spark fears for growth. Accessed 15 October 2015. http://www.ft.com/cms/s/0/d7abbb20-7d65-11df-a0f5-00144feabdc0.html.

Financial Times. (2010d). UK growing near long-term trend, says NIESR. Accessed 30 June 2015. http://www.ft.com/cms/s/0/073c19f8-7590-11df-86c4-00144feabdc0.html.

Foucault, M. (1980). *Power, knowledge: Selected interviews and other writings*. New York: The Harvester Press.

Frowen, S. (Ed.). (1983). *Controlling industrial economies: Essays in honour of Christopher Thomas Saunders*. London: Macmillan.

Guardian. (2008). No recession, economists say—Provided there are no more Rocks. Accessed 20 October 2015. http://www.theguardian.com/business/2008/may/02/creditcrunch.banking.

Guardian. (2009). Revealed: £3bn mistake in George Osborne's budget plan. Accessed 21 October 2015. http://www.theguardian.com/politics/2009/oct/09/george-osborne-budget-deficit.

Guardian. (2013). Romanian and Bulgarian migrants 'unlikely to head for UK'—Report. Accessed 25 October 2015. https://www.theguardian.com/uk/2013/apr/05/eastern-european-migrants-report.

Hierländer, R., Huber, P., Iara, A., Landesmann, M., Nowotny, K., O'Mahony, M., et al. (2010). *Migration, skills and productivity* (WIIW Research Reports, 365). Accessed 20 June 2015. http://wiiw.ac.at/migration-skills-and-productivity-dlp-2270.pdf.

Independent. (1998). Obituary: Christopher Dow. Accessed 20 March 2016. http://www.independent.co.uk/arts-entertainment/obituary-christopher-dow-1189026.html.

Ladi, S. (2011). Think tanks, discursive institutionalism and policy change. In G. Papanagnou (Ed.), *Social science and policy challenges: Democracy, values and capacities*. Paris: UNESCO.

London Review of Books. (2013). *Jonathan Portes: An exercise in scapegoating*. Accessed 10 October 2015. http://www.lrb.co.uk/v35/n12/jonathan-portes/an-exercise-in-scapegoating.

Markoff, J., & Montecinos, V. (1993). The ubiquitous rise of economists. *Journal of Public Policy, 13*(1), 37–68.

McGann, J. (2013). *2012 global go to think tanks and policy advice ranking*. Think Tanks and Civil Society Program, University of Pennsylvania.

Medium. (2015). *Ed Conway: Why Jonathan Portes left NIESR*. Accessed 30 October 2015. https://medium.com/@edconwaysky/why-jonathan-portes-left-niesr-3999329290d9#.t9pn5q6ka.

Medvetz, T. (2012a). *Think tanks in America*. Chicago: University of Chicago Press.

Medvetz, T. (2012b). Murky power: 'Think tanks' as boundary organizations. In D. Golsorkhi, D. Courpasson, & J. Sallaz (Eds.), *Rethinking power in organizations, institutions, and markets: Research in the sociology of organizations* (pp. 113–133). Bingley: Emerald Group Publishing.

Mirowski, P. (2002). *Machine dreams: Economics becomes a cyborg science*. Cambridge: Cambridge University Press.

Norman, J. (2012). *Fisking Portes: How politically neutral *is* the Director of NIESR?* Accessed 23 October 2015. http://www.jessenorman.com/2012/11/fisking-portes-how-politically-neutral-is-the-director-of-niesr.html.

Norpoth, H. (1991). The popularity of the Thatcher government: A matter of war and economy. In H. Norpoth, M. Lewis-Beck, & J. Lafay (Eds.), *Economics and politics: The calculus of support*. Ann Arbor: University of Michigan Press.

Pain, N., & Young, G. (2004). The macroeconomic impact of UK withdrawal from the EU. *Economic Modelling, 21*, 387–408.

ParliamentLive. (2012). *Treasury Committee: Tuesday 13 November 2012*. Accessed 10 October 2015. http://www.parliamentlive.tv/Event/Index/a31b4072-101e-4a65-8a90-f14cba88937d.

Pielke, R. (2007). *The honest broker: Making sense of science in policy and politics*. Cambridge: Cambridge University Press.

Pieria. (2013). *Jonathan Portes: NIESR and the European Commission: A response to Dan Hannan*. Accessed 12 October 2015. http://www.pieria.co.uk/articles/niesr_and_the_european_commission_a_response_to_dan_hannan.

Porter, T. (1995). *Trust in numbers: The pursuit of objectivity in science and public life*. Princeton, NJ: Princeton University Press.

Prospect. (2013). The Prospect duel: Was Osborne right? Accessed 10 October 2015. http://www.prospectmagazine.co.uk/regulars/the-prospect-duel-mark-littlewood-jonathan-portes-was-osborne-right.

Reichmann, W. (2011). Institutionalizing scientific knowledge: The social and political foundation of empirical economic research. *Sociology Compass, 5*(7), 564–575.

Schmidt, V. (2008). Discursive institutionalism: The explanatory power of ideas and discourse. *Political Science, 11*(1), 303–322.

The Spectator. (2015a). Niall Ferguson: Jonathan Portes, master of the political correction. Accessed 10 October 2015. http://www.spectator.co.uk/2015/06/jonathan-portes-master-of-correct-politicalness/.

The Spectator. (2015b). Jonathan Portes fired as NIESR director. Accessed 30 October 2015 http://blogs.spectator.co.uk/2015/10/jonathan-portes-stands-down-from-niesr-by-mutual-consent/.

The Times. (2004). Brown attacked for fiscal 'fudge.' Accessed 24 October 2015. http://www.thetimes.co.uk/tto/business/entrepreneur/article2229114.ece.

Weale, M. (2010). Unsustainable consumption: The structural flaw behind the UK's long boom. In G. Giudice, R. Kuenzel, & T. Springbett (Eds.), *UK economy: The crisis in perspective* (pp. 59–79). London: Routledge.

Weingart, P. (1999). Scientific expertise and political accountability: Paradoxes of science in politics. *Science and Public Policy, 26*(3), 151–161.

Think Tank Reports and Blog Posts [NIESR, available at niesr.ac.uk]

Armstrong, A. (10/2011). An assessment of the Independent Commission on Banking's recommendations. *National Institute Economic Review, 218*(3), f4–f12.

Babecký, J., Bulíř, A., & Šmídková, K. (04/2008). Sustainable real exchange rates when trade winds are plentiful. *National Institute Economic Review, 204*, 98–107.

Bagaria, N., Holland, D., & Van Reenen, J. (07/2012). Fiscal consolidation during a depression. *National Institute Economic Review, 221*, f42–f54.

Barrell, R. (10/2008). *Press release: The great crash of 2008.*

Barrell, R. (10/2009a). Budget consolidation options for the UK. *National Institute Economic Review, 210*, 58–60.

Barrell, R. (10/2009b). Long-term scarring from the financial crisis. *National Institute Economic Review, 210*, 36–38.

Barrell, R., & Davis, P. (04/2011). Financial regulation. *National Institute Economic Review, 216*, F4–F9.

Barrell, R., Foley-Fischer & Kirby, S. (10/2009). Prospects for the UK economy. *National Institute Economic Review, 210*, 39–57.

Barrell, R., & Hurst, I. (04/2008). Financial crises and the prospects for recession. *National Institute Economic Review, 204*, 33–38.

Barrell, R., & Kirby, S. (04/2007). Managing a decline in inflation. *National Institute Economic Review, 200*, 53–55.

Barrell, R., Kirby, S., Metz, R., & Weale, M. (04/2005). The Labour government's economic record and economic prospects. *National Institute Economic Review, 192*, 4–10.

Barrell, R., & Weale, M. (03/2009). *The economics of a reduction in VAT* (NIESR Discussion Paper, 325).

Budd, A. (04/2010) Fiscal policy under Labour. *National Institute Economic Review, 212*, R34-R48.

Dorsett, R., Lui, S., & Weale, M. (08/2010). *Economic benefits of lifelong learning.*

Goodridge, P., Haskel, J., & Wallis, G. (05/2013). Can intangible investment explain the UK productivity puzzle? *National Institute Economic Review, 224*, f48–f58.

Holland, D., & Portes, J. (10/2012). Self-defeating austerity? *National Institute Economic Review, 222*, f4–f10.

Kirby, S., Paluchowski, P., & Warren, J. (05/2014). The performance of NIESR's GDP forecasts. *National Institute Economic Review, 228*, 66–70.

Kirby, S., & Riley, R. (01/2007). UK economy forecast. *National Institute Economic Review, 199*, 40–58.

Jones, K. (05/1988). Fifty years of economic research: A brief history of the National Institute for Economic and Social Research. *National Institute Economic Review, 124*, 36–62.

Jones, K. (04/1998). *Sixty years of economic research: A brief history of the National Institute of Economic and Social Research*. London: NIESR Occasional Papers.

Metcalf, H., & Rolfe, H. (12/2010). *Caste discrimination and harassment in Great Britain.*

NIESR. (08/2007). *2007 annual report.*

NIESR. (01/2008). The world economy: Risks of a global recession. *National Institute Economic Review, 203*, 8–30.

NIESR. (07/2008). The UK economy. *National Institute Economic Review, 205*, 3.

NIESR. (08/2008). *2008 annual report.*

NIESR. (01/2009). *Estimates of monthly GDP.*

NIESR. (04/2009). The UK economy. *National Institute Economic Review, 208*, p. 3.

NIESR. (08/2009). *2009 annual review.*

NIESR. (11/2009). *The channels of financial market contagion in the new EU member states.*

NIESR. (03/2011). *2011 annual accounts.*

NIESR. (01/2012). The UK economy. *National Institute Economic Review, 219*, f3.

NIESR. (03/2012). *2012 annual accounts.*

NIESR. (07/2012). The UK economy. *National Institute Economic Review, 221*, f3.

NIESR. (02/2013). The UK economy. *National Institute Economic Review, 223*, f3.

NIESR. (08/2013). The UK economy. *National Institute Economic Review, 225*, f3.

NIESR. (05/2014). *Prospects for the UK economy.*

Portes, J. (01/2012a). *British jobs and foreign workers: Today's reports on immigration and unemployment.*

Portes, J. (01/2012b). *Fiscal policy: What does "Keynesian" mean?*

Portes, J. (03/2012). *It's not too late to change course: Macbeth and fiscal policy.*
Portes, J. (04/2012). *The European Commission is doing its best to drive Spain to disaster.*
Portes, J. (06/2012a). *DWP analysis shows mandatory work activity is largely ineffective. Government is therefore extending it.*
Portes, J. (06/2012b). *Macroeconomics: What is it good for? [a response to Diane Coyle].*
Portes, J. (10/2012). *More on multipliers: Why does it matter?*
Portes, J. (11/2012). *Faith-based economics at the Treasury Committee.*
Portes, J. (01/2013). *Debating debt ratios with Michael Fabricant.*
Portes, J. (03/2013). *Budget 2013: Time for an investment-led growth agenda.*
Portes, J. (10/2013). *Fiscal consolidation and growth: What's going on?*
Portes, J. (11/2013). Commentary: The economic implications for the UK of leaving the European Union. *National Institute Economic Review, 226*, 4–9.
Riley, R., & Young, G. (05/2014). Financial crisis and economic performance: Introduction. *National Institute Economic Review, 228*, r1–r2.
Robinson, A. (05/1988). The National Institute: The early years. *National Institute Economic Review, 124*, 63–66.
Rolfe, H., Fic, T., Lalani, M., Roman, M., Prohaska, M., & Doudeva, L. (04/2013). Potential impacts on the UK of future migration from Bulgaria and Romania.
Weale, M. (10/2007). Northern Rock: Solutions and problems. *National Institute Economic Review, 202*, 4–8.
Weale, M. (04/2008). Commentary: The banking crisis and the economy. *National Institute Economic Review, 204*, 4–8.
Weale, M. (10/2008). Commentary: The burden of the national debt. *National Institute Economic Review, 2010*, 4–8.

6

Policy Exchange: The Pros and Cons of Political Centrality

In the early 2000s, Conservative MPs Francis Maude and Archie Norman established two organisations to prop up their party's appeal after the rise of New Labour. Attempting to move the Conservatives towards the centre-ground, they were following the footsteps of Michael Portillo's unsuccessful leadership bid in the aftermath of the electoral defeat of the 2001 General Election (*Daily Telegraph* 2001). The first of these organisations was a pressure group, C-Change, whose purpose was fulfilled after the ascent of Conservative modernisers when David Cameron became leader, after yet another defeat against New Labour in 2005 (Maude 03/2012). The second was a think tank, X-Change ideas, later rebranded Policy Exchange (PX).[1] A formally independent, non-party political charity, PX's mission was to develop ambitious policy proposals to renew centre-right thinking. Behind PX's foundation was the diagnosis that the British right and

[1]See UK Companies House, Reg. No. 04297905, accessed 1 March 2016, https://beta.companieshouse.gov.uk/company/04297905/filing-history?page=5.

M. González Hernando, *British Think Tanks After the 2008 Global Financial Crisis*, Palgrave Studies in Science, Knowledge and Policy, https://doi.org/10.1007/978-3-030-20370-2_6

its associated think tanks (IEA, CPS, ASI) was overburdened by its history and had become less agile than its counterpart, which under Tony Blair propounded a 'post-ideological' approach to policymaking (Denham and Garnett 2006).

PX thus became associated with the moderniser wing of the Conservatives, which believed "political recovery was reliant on the party acclimatising to the changing social and political landscape and adopting a much more liberal line on 'lifestyle' issues, [...] embracing much of the 'modern' Blairite social policy agenda" (Williams 2015: 10–11). Hence, PX could be considered a mirror image of third-wave centre-left think tanks such as IPPR and Demos, which were themselves a response to 1970s new-right think tanks (Denham and Garnett 1998). Because of the above, PX's foundation cannot be historically dissociated from the pragmatic 'what works' discourse of the New Labour years (Denham and Garnett 2006; Schlesinger 2009). Yet, in opposition to what they saw as the target-led, top-down management style of Blair and Brown, PX would put forward evidence-based policy solutions with an accent on free markets, voluntarism, and localism, bringing decision-making away from central government to local communities (Jenkins 11/2004).

As modernisers rose in Conservative ranks, PX became ever more entwined in networks of political power and, thanks to its many connections in high places, came to be considered among the most influential think tanks in the UK. So much so that in 2005, future Prime Minister David Cameron delivered a speech in the run-up to the Conservative leadership contest in PX's premises. In that speech, in line with PX's efforts to move the centre-right beyond the legacy of Thatcherism, Cameron stated: "[w]e do think there's such a thing as society, we just don't think it's the same thing as the state" (*Guardian* 2005).

Sometimes labelled 'Cameron's favourite think tank' (*Financial Times* 2008a), PX has employed many future Conservative politicians and *apparatchiks*. Nick Boles (later MP) was its first Director (2002–2007), followed by Anthony Browne (2007–2008) (later Mayor of London's Policy Director) and Neil O'Brien (2008–2013) (later Special Adviser to the Treasury and MP). Many of PX's current and past members are also prominent figures at the interface of journalism and politics. Such cases include former *The Spectator* Associate Editor and current PX

Director Dean Godson,[2] and PX's successive Chairmen of the Board of Trustees: *Times* columnist and future Minister Michael Gove[3] (2002–2005), former *Daily Telegraph* Editor Charles Moore[4] (2005–2011), former *Times* Executive Director Daniel Finkelstein[5] (2011–2014), and, across the pond, *The Atlantic* Senior Editor David Frum.[6] To add further examples of PX's political connections: future Conservative MPs Jesse Norman,[7] Charlotte Leslie, and Chris Skidmore were PX employees—Norman was Executive Director (2005–2006), Leslie and Skidmore were Research Fellows. The future Policy Director for the Prime Minister's Office James O'Shaughnessy[8] was Deputy Director (2004–2007), followed in the same position by now Baroness Natalie Evans[9] (2008–2011). Also, DEFRA Special Advisers Amy Fischer and Ruth Porter were, respectively, Head of Communications (2008–2010) and Head of 'Economics & Social Policy' (2014).

Given these connections to the Conservative party, PX has been considered a 'revolving door,' a recruitment ground for "the party's best" (Pautz 2012: 8–9), a factory of suitable policy proposals and, perhaps more importantly, the crucible of new ideas for the centre-right's policy-making elite (*Financial Times* 2008b). As such, particularly after 2005, PX members partook in the formulation of the Conservative platform in succeeding elections. Indeed, the shift from a discourse narrowly

---

[2]Dean Godson joined PX in 2005, and was previously Chief Leader Writer of *The Daily Telegraph*, Associate Editor of *The Spectator*, and Contributing Editor for *Prospect Magazine*.

[3]Michael Gove became an MP in 2005 and later Education, Justice, and Environment Secretary.

[4]Charles Moore is former editor of *The Daily Telegraph*, *The Sunday Telegraph*, and *The Spectator*. After leaving his post at PX, Moore wrote the authorised biography of Margaret Thatcher.

[5]Daniel Finkelstein is former Director of the think tank Social Market Foundation and became Member of the House of Lords in 2013.

[6]Daniel Frum was a speechwriter for US President George W. Bush and Chairman of American Friends of PX.

[7]As reported in Chapter V, Jesse Norman is also member of NIESR's Board of Trustees.

[8]James O'Shaughnessy co-authored the Coalition's programme, was "Director of the Conservative Research Department from 2007 and 2010, and helped write the Conservative Party's general election manifesto," accessed 3 March 2016. http://www.policyexchange.org.uk/people/alumni/item/james-o-shaughnessy?category_id=45.

[9]Natalie Evans is also former Head of Operations at the New Schools Network, a Charity supporting new independent schools.

focused on free markets towards pushing for a 'Big Society' that prioritised local communities and voluntary associations was partly influenced by PX publications (Norman 12/2008). Parallel to its ideological role, PX has arguably functioned as a 'garbage can' for Conservative policies (Stone 2007), producing a battery of proposals to be implemented when a 'window of opportunity' opened, especially in areas hitherto unfavoured by the Conservatives' focus on fiscal discipline, such as education, community relations, the environment, etc.

Although it does not perform as highly as one would expect in the University of Pennsylvania's *Global go-to* think tank ranking,[10] PX has a strong reputation for political impact (Pautz 2012). In *Prospect's think tank awards*, PX earned the 'one-to-watch' prize in 2004, best publication in 2005—for *Unaffordable housing* (Evans and Hartwich 06/2005)—and 'think tank of the year' in 2006. Its own webpage states plainly "Policy Exchange is the UK's leading think tank."[11]

Organisational and Funding Structure

PX's offices are in Clutha House, an august building where several policy-related organisations are also located, including Civitas (previously occupied by its Centre for Social Cohesion), the Pilgrim Trust, and Localis, a local government think tank with several ties to PX.[12] The premises are across the street from the Houses of Parliament, and

[10]Between 2012 and 2015, PX appears on these rankings in the following positions: World think tanks (non-US): 95, 95, 95; 113; World think tanks (US and non-US): 116, 118, n/a; n/a; Think tanks in Western Europe: 70, 71, 75, 73 (McGann, 2013–2016).

[11]Accessed 20 March 2016, https://web.archive.org/web/20121011031859/http://policyexchange.org.uk/about-us.

[12]Anthony Browne, Neil O'Brien, and Nicholas Boles were all members of Localis' board (data retrieved from Companies House, Localis Research Ltd. Reg. No. 04287449), accessed 30 March 2016, https://beta.companieshouse.gov.uk/company/04287449/filing-history.

their first floor accommodates the 'Ideas Space,'[13] the name of both a subsidiary company in charge of organising PX's events and the venue where PX holds the majority of them.

PX's propinquity to Westminster and centre-right circles partly explains the significant attention its research accrues, especially in the lead-up to an expected Conservative victory in the 2010 general election. As such, PX permanent staff has expanded considerably, from 22 in 2007 to 35 in 2013. Of these employees, roughly twenty are researchers, while the rest work on administration, communications, events, and fundraising. Though PX is a sizeable think tank, it is far from the largest in Britain—it is smaller, for instance, than NEF and NIESR. Regardless, PX publishes the most policy reports among our sample, 251 between 2007 and 2013, many by fellows and external authors. Being a generalist organisation, its work spans a wide gamut of policy areas, covering mainstream—e.g., deficit reduction (Holmes et al. 04/2010)—abstract—the meaning of 'fairness' (Lilico 02/2011)—and 'niche' issues—prisoner electronic monitoring (Geoghegan 09/2012). The topics PX studies are organised in several research units. Within our timeframe, these are 'Crime & Justice,' 'Digital Government,' 'Economics & Social Policy,' 'Financial Policy', 'Housing & Planning,' 'Education & Arts,' 'Environment & Energy,' 'Government & Politics,' 'Health' and 'Security,' to which a 'Demography, Immigration and Integration' unit was added in 2015.

Organisationally, this think tank's most distinctive characteristic is its many links to Conservative politicians, donors, and sympathisers. Although as a charity PX is formally party-independent, its closeness to Tory modernisers is certainly part of its brand. And although comparable think tanks on the centre-right also sprung around the time of PX's foundation—notably Civitas, Reform, and the Centre for Social Justice (CSJ)—few have their perceived closeness to Cameron's premiership. To give one illustration of PX's position within the mainstream

[13]Now 'Policy Exchange Events.' See UK Companies House, Reg. No. 06005752, accessed 1 March 2016, https://beta.companieshouse.gov.uk/company/06005752/filing-history.

of British politics, its board of trustees includes or has included several party donors and members of the right-wing press. Among them we count Tory fellow travellers—Rachel Whetstone,[14] Camilla Cavendish,[15] Richard Ehrman,[16] Patience Wheatcroft[17]—journalists—Virgina Fraser,[18] Alice Thompson[19]—and donors and businesspeople—Theodore Agnew,[20] Richard Briance,[21] Simon Brocklebank-Fowler,[22] David Meller,[23] George Robinson,[24] Robert Rosenkranz,[25] Andrew Sells,[26] Tim Steel,[27] and Simon Wolfson.[28] Links to Conservatives also take more symbolic forms: in 2010, PX held its first annual lecture in memory of Conservative businessman and philanthropist Leonard Steinberg, and in 2011 another such lecture was inaugurated commemorating Lord Christopher Kingsland, late Tory peer, and MEP.

PX's political connections are so conspicuous that in 2009 the *Daily Telegraph* (2009) counted six of its members among the one hundred most influential right-wingers in Britain—a list in which no ASI staff

[14]Rachel Whetstone is the former Head of Communications and Public Policy at Google, was political secretary to former Conservative leader Michael Howard, and is married to Steve Hilton, David Cameron's speechwriter and visiting scholar at PX.

[15]Camilla Cavendish headed the 2013 NHS 'Cavendish Review' and later became Head of the Prime Minister's Office Policy Unit.

[16]At the time of writing, Richard Ehrman was Consultant Director at the think tank Politeia.

[17]Patience Wheatcroft is Conservative life peer and former editor of *The Wall Street Journal Europe*.

[18]Virginia Fraser writes for *Homes & Gardens* magazine and has worked as an editor for *The Sunday Telegraph* and *The Spectator*.

[19]Alice Thompson is Associate Editor at *The Times*.

[20]Theodore Agnew is a Conservative donor and later non-executive board member of the Department of Education and Head of Academies. See (accessed 1 March 2016) https://www.gov.uk/government/people/theodore-agnew#previous-roles.

[21]Richard Briance is Partner of PMB Capital.

[22]Simon Brocklebank-Fowler is the founder of Cubitt Consulting.

[23]David Meller is Chair of the Meller Group and founder of the Meller Education Trust.

[24]George Robinson is Director of the hedge fund Sloane Robinson.

[25]Robert Rosenkranz is CEO of Delphi Financial Group and member of the Yale University Council, the Council of Foreign Relations, and the board of the Manhattan Institute.

[26]Andrew Sells is Chairman of the non-departmental public body Natural England, accessed 25 March 2016, https://www.gov.uk/government/people/andrew-sells.

[27]Tim Steel is Chairman of the private equity firm Committed Capital.

[28]Baron Simon Wolfson is the CEO of the clothing retailer Next and a Conservative life peer.

member appears. Internationally, PX was a part, while it existed, of the free-market Stockholm Network and, in 2010, American Friends of Policy Exchange (AFPX) was established. Spearheaded by David Frum—also PX trustee chairman—AFPX has 501(c)3 non-profit status, which allows it to channel funding from US donors.

Given their social and spatial proximity to centres of power, PX is in a privileged position to host public events. For that reason, their function as a 'venue' should be considered central to their brand as a think tank. Besides the launching of policy reports, PX events take several formats. One key type—on account of the 'political capital' they garner and the press coverage they receive—is set speeches by politicians, which PX hosts even in areas not directly covered by their research.[29] A good illustration of this is that both the former Home Secretary and former Immigration Minister have spoken on immigration reform at 'The Ideas Space' (Gov.uk 2010a, 2012a), a policy issue on which PX published very little before 2015. Other relevant events formats PX caters for include talks by prominent experts and academics (e.g., economist John Kay, social capital theorist Robert Putnam, behavioural economist Richard Thaler), roundtables, and debates—dubbed 'Policy Fight Clubs.'

In terms of the provenance of guest speakers, even if PX is most strongly linked to the centre-right, politicians from across the spectrum, both from the UK and overseas, have also been invited to 'the Ideas Space.'[30] Figures of the prominence of Wolfgang Schäuble (German Interior Minister), General David Petraeus (former CIA Director), Frank Luntz (Republican Party strategist), and former Prime Ministers José María Aznar from Spain and John Howard from Australia have spoken at PX. Among British speakers, Conservative politicians are

---

[29]"[W]e're known […] as a good space for discussion […]. So if a minister or a shadow minister […] wants to make a policy announcement, we will almost certainly try to help them" (PX interview).

[30]Indeed, much publicity to PX comes in the form of speeches by senior figures at their premises, and one finds that copyright-free images of speakers with the PX logo as background are available around the web—e.g., Education Ministers Michael Gove and Nicky Morgan's Wikipedia profiles. See (accessed 20 March 2016) https://en.wikipedia.org/wiki/Michael_Gove and https://en.wikipedia.org/wiki/Nicky_Morgan.

most common—both 'frontline' and 'backbenchers,' modernisers and otherwise. However, in 2010 PX hosted Ed Miliband (then bidding for the Labour party leadership). Labour MPs Jon Cruddas and Diane Abbott have also spoken at 'The Ideas Space,' as well as Liberal Democrats Vince Cable and Chris Huhne, to name but a few. In line with this ecumenical approach, PX is routinely present in Labour, Liberal Democrat, and Conservative Party conferences. Given the potential audience of their events, it is perhaps unsurprising that many have been funded or co-hosted by major private organisations (Deloitte, Google, Microsoft, PwC, Serco) and educational and charitable trusts (Cambridge Assessment, Esmee Fairbairn Foundation, New Schools Network, Sutton Trust, TeachFirst). In relation to why supporting PX events is an attractive proposition, one interviewee said:

> Mostly I would say [funders get] two things. One to be associated with the reform of the welfare system or whatever it might be to promote their brand, [...] and then the other thing is that they get to [...] say 'this is our event' and they get nice seats, they get to meet the minister or an important person, they get to do the press release, put their posters up [...]. And frankly [...] I think seeing it from the other side its worthwhile to them because, you know, for a huge company like PwC or something, think tanks are incredibly small and sponsorship of an event is a tiny amount of money for them. (PX interview)

Financially, PX has had a mostly favourable situation between 2007 and 2013, bolstered by close relationships with business and Conservative donors and a reputation for policy impact. Figures available at the Charity Commission are reported Table 6.1 which showcase an impressive yearly growth. Its total income has risen from under £1m in 2006 to over £2.5m in 2013 and runs a significant amount of reserves. However, according to the criteria of two think tank transparency surveys—Whofundsyou[31] and Transparify (2016)—PX funding is

[31]Accessed 20 March 2016, http://whofundsyou.org.

Table 6.1 PX financial overview

PX funding (£)	2007	2008	2009	2010	2011	2012	2013
Restricted income	1,280,803	1,101,274	1,779,996	1,115,452	1,223,923	2,377,972	1,616,473
Unrestricted	506,022	1,557,693	915,830	961,368	947,517	846,190	903,442
Total income	1,786,825	2,658,967	2,695,826	2,076,820	2,171,440	3,224,162	2,519,915
% Unrestricted	28.3	58.5	33.9	46.2	43.6	26.2	35.8
Total expenses	1,585,411	2,188,770	2,204,807	2,148,827	2,170,296	2,898,384	2,491,860
Balance	201,414	470,173	491,019	−72,007	1144	325,778	28,055

Source Data from financial statements supplied to the UK Government Charity Commission (2008–2013, Charity No: 1096300), accessed 20 March 2016, http://apps.charitycommission.gov.uk/Showcharity/RegisterOfCharities/DocumentList.aspx?RegisteredCharityNumber=1096300&SubsidiaryNumber=0&DocType=AccountList

secretive, on account of the lack of a centralised list of contributions. Nevertheless, although British law does not demand think tanks to disclose their sources, donors who so wish are mentioned in PX reports.

PX funding can be divided, in the words of an interviewee, in "roughly a third from individual donors, a third from trust and foundations, and a third from corporate sponsorship" (PX interview).[32] Contributions by individual are difficult to trace, but PX reports to have been funded by trusts belonging to Conservative donors such as the Law Family Charitable Trust[33] and The Peter Cruddas Foundation. The latter donated £140,000 in 2008[34] to research public service delivery for disadvantaged groups, plus £300,000 in 2009[35] and £120,000 in 2010[36] for work on child poverty, broken families, and supply side reforms. Other charitable foundations who fund PX include the Barrow-Cadbury Trust, which has sponsored research on older workers (Tinsley 06/2012), the Hadley Trust on criminal justice reform, AgeUK on care for the elderly (Featherstone and Whitham 07/2010), plus educational trusts such as the Sutton Trust, Cambridge Assessment, and the Edge Foundation (Davies and Lim 03/2008). In terms of corporate contributions, many of them are raised through the PX Business Forum, which has enticed companies such as Reliance plc, BSkyB, Terra Firma, SAB Miller, and Bupa. To these one could add funding from trade associations such as BVCA, CCIA, and The City of London Corporation. In terms of what type of funding PX does not receive compared to other case-studies, academic research councils come to mind.

[32]"So the funding structure is roughly the corporate sponsors, [...] contribute to an annual membership fee like a forum, a business forum. So they underwrite a lot of the operational costs if you will. And then there's events which are paid for, it could be anyone, mostly corporates, charities sometimes. And then we fund individual pieces of research. Sometimes that's through foundations like Joseph Rowntree for example [and] there are very rich donors who like what we do and will contribute" (PX interview).

[33]Accessed 30 March 2016, http://www.lawfamilycharitablefoundation.org.

[34]Accessed 21 February 2016, http://www.petercruddasfoundation.org.uk/docs/Annual-Report-and-Financial-Statements-year-ended-31-3-08.pdf.

[35]Accessed 21 February 2016, http://www.petercruddasfoundation.org.uk/docs/Annual-Report-and-Financial-Statements-year-ended-31-3-09.pdf.

[36]Accessed 21 February 2016, http://www.petercruddasfoundation.org.uk/docs/Annual-Report-and-Financial-Statements-year-ended-31-3-10.pdf.

Overall, PX's growing income and the breadth of its funders result in robust finances and reserves that allow for some leeway in the type of research it can pursue. This relative affluence granted a degree of responsiveness that is sometimes lacking in most think tanks that rely on commission funding, as is the case with NEF and NIESR, while having sufficient resources to produce long-term in-house empirical research, unlike the ASI.[37] One interviewee claimed: "if the director decides I really ought to look into this area, then it will happen, even if there has to be some funding centrally" (PX interview). An instance of this is visible in the PX annual accounts of 2009–2010, which show 'transfers from designated reserves' of £70,000 to the Economics and £50,000 to the Education units—perhaps an index of shifting research and policy priorities. For the above reasons, since PX was closely linked to a Conservative Party that was all but set to become government after 17 years, this think tank had a privileged position to propose a route of travel for a new administration. However, this closeness to political elites could elicit tensions, as will be shown apropos the interplay between the 'Big Society' discourse and austerity.

Style and Tropes

I shall start with their mission statement: "Policy Exchange is an independent, non-partisan educational charity seeking free-market and localist solutions to public policy questions. [...] The authority and credibility of our research is our greatest asset."[38] Not unlike other think tanks, PX's self-definition refers to their normative orientation—in their case towards free markets and localism—and their adherence to rigorous research. To advance their values in consideration of empirical evidence, PX's goal is to propose innovative yet reasonable and politically viable

[37]"We had problems funding certain things. But we never [...] really had a huge amount of money shortfall. So we were able to be very independent for that reason" (PX interview).

[38]Accessed 20 March 2016, https://web.archive.org/web/20121011031859/http://policyexchange.org.uk/about-us.

policies. According to former deputy director James O'Shaugnessy,[39] PX—unlike many other British think tanks, which advocate for overly theoretical or impracticable ideas—sets itself apart by producing 'off the shelf' policy proposals that could readily be used as blueprints for ministers to implement.

However, these characteristics are not specific enough to describe what makes PX distinct, as most think tanks declare a commitment to empirical evidence and an orientation towards practical policy proposals animated by some set of values. What is peculiar of PX is its historical origin among Conservative modernisers in a context of electoral defeat and ideological renewal and, therefore, its closeness to the Conservative leadership after 2005. PX was born as part of an attempt to deal with the weaknesses of the British right an era when 'Blairism' held sway. At that time, in the words of a commentator, "the most significant division in the Conservative Party was along the social, sexual, and moral policy divide" (Hayton 2012: 117). In other words, although most Tories agreed on the issue of free markets, modernisers put a greater accent on social justice and inclusiveness. They also were, at least in principle, willing to distance themselves from the more rigid approach to welfare of politicians and think tanks on their right. Hence PX's principal focus would be, rather than on fiscal policy, in 'softer' areas like education, policing, housing, and security.

Arguably, the modernisers behind PX's foundation were the mirror image of New Labour, which, opposed to what they saw as a more ideological 'Old' Labour, advanced a political platform centred on 'what works' and 'evidence-based policymaking.'[40] However, PX was otherwise critical of New Labour, as they believed was too top-down, managerial, and target-led, curtailing the potential of individuals and communities rather than harnessing it. In contrast, PX would propose pragmatic and locally based solutions to policy problems, centred on localism, social justice, and voluntary and local associations. In line with this push for renewal, most research by PX was traditionally outside narrowly defined economic or fiscal issues, towards areas Conservative

[39]See 11:50 mark, accessed 7 March 2016, https://www.youtube.com/watch?v=RAASzMcKZ_w.

[40]See former PX Director Neil O'Brien in the Labour conference fringe event 'Reconsidering Blairism,' accessed 7 March 2016, https://www.youtube.com/watch?v=xVVAX-RPEvY.

thinking had mostly neglected. In the words of an interviewee, PX's "original *raison d'être* is to produce some Conservative ideas that aren't about tax, Europe, or the deficit" (PX interview).

Although PX defines itself as pro-free markets, as a rule, its public interventions are less categorical than those of, in the words of a former employee, the free market's Praetorian guard (CPS, ASI, IEA). In Maude's (03/2012) narrative, it was PX very purpose to provide Conservatives with middle-ground ideas that went beyond 1980s-style Monetarism. That has meant an emphasis on practical policy problems and an awareness of where the political centre of gravity is at any given time, rather than intellectual stability and seeking to pull the Overton window from one end. This pragmatic orientation sometimes generated a certain distance from more unflinching sectors of the right. Two examples of this are their use of behavioural economics and their work on environmental policy. On the first, PX has embraced across policy areas the findings of *Nudge* (Sunstein and Thaler 2008), which advocates for policies that, without curtailing individuals' rights, make socially efficient outcomes more likely through schemes informed by behavioural economics. These ideas were argued even if, given *Nudge*'s inherent 'paternalism,' they have met apprehensions from free-market advocates (ConservativeHome 2010). On the second, the environment—an area in which the 'nudge' approach has been applied extensively (Newey 07/2011, 01/2013)—rather than contesting the very existence of anthropogenic climate change like other sectors of the right, PX has aimed to devise localist and market-based solutions to tackle it (Helm 12/2008; Drayson 11/2013).

Since its foundation, PX has also sought to redefine what Conservatism means, partly to widen its electoral appeal—engaging in 'boundary work,' as it were. For instance, against sectors of the Conservative Party base, PX has advocated that gay marriage is a fundamentally conservative policy and will strengthen rather than weaken the institutions of family and marriage (Flint and Skelton 2012). This is consistent with the Coalition's policy agenda, as it approved same-sex marriage legislation in England and Wales in 2013. In summary, PX has sought to brand itself and Conservatism as practical, flexible, and inclusive. PX founder Francis Maude has even claimed his party, unlike their main political opponents, is not animated by 'dogma,' but open to change with changing times. In his own words:

> Like British society, the Conservative Party has the suppleness that allows it to adapt to a new moment, a new time and to absorb the spirit of the age. Not because we lack conviction but because unlike the Labour Party, the Conservative Party was never a party of dogma. (Maude 03/2012)

PX has also published pamphlets and books work setting out the philosophy of Tory modernisation. Such publications include those of FT columnist Janan Ganesh and Jesse Norman MP (06/2006) on 'Compassionate Conservatism' and "The Big Society" (see also Norman 12/2008, 2010) which sought to bring together much of PX's ideas across policy areas through an overarching narrative. In that corpus, Norman and Ganesh argue that twentieth-century politics were defined by the conflict between those who argued for a greater role for the state or for the market. In the process, a considerable area of human concerns, civil society, had often remained unnoticed. The mission of a future conservatism should be, instead of re-enacting past debates on the remit of the public and private sector, to foster the development of a robust civil society by supporting local communities, charities, and independent associations. This commitment to social cohesion is even visible in reports under the 'security' banner, as PX has argued to 'stop emphasising difference' to prevent radicalisation among British Muslims (Mirza et al. 01/2007). In sum, PX's ideas helped set the background for the Conservative's 2010 electoral platform, which at least until 2008 was centred on promoting the 'Big Society.' This broad notion, and its focus on mending a 'Broken Britain' (Evans 2011), was Cameron's 'vehicular idea' (McLennan 2004), much like the 'Third Way' had been for Blair.

There are two further common tropes in PX public interventions that are worth noting here. The first is their assertion—in a notice present in most policy reports (e.g., Evans and Hartwich 01/2007: 2)—that "government has much to learn from business and the voluntary sector." Connected to moderniser's aim to strengthen local associations, PX argues for a leaner state and a stronger third sector, for instance, by encouraging philanthropy (Davies and Mitchell 11/2008). The second trope is also present in the aforementioned notice: "[w]e believe that the policy experience of other countries offers important lessons for government in the UK." In line with their orientation towards producing workable policies, PX often imports what they see as successful policy ideas from other countries. The areas in which PX has drawn from international and historical experience include counterterrorism (Reid 03/2005), fiscal consolidation (Holmes et al. 11/2009), and education (Davies and Lim 03/2008).

Yet another distinctive characteristic of PX's public interventions concerns their policy focus. Possibly aided by their access to senior political figures and their keen awareness of the internal debates within Conservative circles, PX has arguably a more impact-focused approach to their research agenda than most think tanks. Rather than being primarily guided by available research-commission funding, PX's agenda centres on areas where they are likely to have the most influence. That is, PX aims to produce timely proposals that are politically workable and which governments are plausibly able and willing to implement. In Kingdon's (2003) terms, PX is especially mindful of policy 'windows of opportunity.' On this point, the words of one interviewee are worth quoting at length:

> One of the important things for a think tank is to be relevant [...] at the end of the day what you're writing is just a piece of paper if no one reads it, so if it's not relevant to the political narrative, if political parties aren't looking seriously at changing these things, or if they've already decided what to do then there's not much point in saying let's look at it. So, for example [...] we did have a Health unit that was looking at NHS reform [...] and we [...] abandoned that programme [...]. Why? Because the government abandoned most of the reforms and made it clear they were not for big changes, so there wasn't much point. (PX interview)

Hence, as their mission is influencing policy in practice, the main target audience for PX's work is the policymaking elite, rather than the wider public debate or any of its subgroups. Engagement with the media, albeit important, is seen by interviewees as secondary to seeking policy change through reaching civil servants and politicians. In their words:

> [...] it certainly helps to be on the television screen or to have a byline in a newspaper and part of trying to change policy is not just going and seeing a minister and trying to change their mind. It's also about changing the public discourse [...] But when it comes to what we're about [...] is trying to change policy in government or in political parties who might get into government [...] And for that [...] you're trying to persuade a completely different audience, you're not talking to the public really, you're talking to half a dozen people, senior civil servants, special advisers, and the minister, and that requires a completely different tone. (PX interview)

Furthermore, even though PX does produce some media commentary and their reports often reach larger audiences through its coverage by broadsheets and the broadcast appearances of its members, interviewees believe media visibility has its risks. For a think tank that has sought to define itself as moderate and centre-right, the media's common framing of issues along antagonistic lines might elicit pressures to take a categorical position they are sometimes uncomfortable with. One interviewee said (example redacted):

> [Media organisations] want not just someone to speak as an expert on a topic, they want someone with particular view [...] and if they don't like what they hear then they will not take you up [...] So sometimes you've got to resist the temptation to say what they want you to say if you don't agree. [...] That might [be] good for me, but it wouldn't be good for Policy Exchange or my reputation. It's not uncommon though. It's a balancing thing. [The media] wants a debate, they want someone on one side and someone on the other. And you're more likely to get requests if you're bang on one side and bang on the other. I've [...] been told that 'you're literally too balanced'. (PX interview)

The Policy Shop at a Time of High Demand

While expanding on what I meant by *hysteresis*, I ventured that a think tank with close connections to a party expected to become government would probably generate policy proposals to establish a clear route of travel. However, in this particular case, being that the official

Conservative policy response to the financial crisis in Britain was fiscal consolidation, this could generate tensions between PX's advocacy for a 'Compassionate Conservatism' and a narrative on the need for fiscal discipline that was becoming ever more pervasive.

This tension, to be sure, assumes that PX remained close to Cameron's leadership. However, two PX reports sparked controversies that threatened this position. The first, *The hijacking of British Islam* (MacEoin 10/2007) claimed dangerous literature was being sold in British mosques. Soon later, BBC reporters accused PX of forging evidence (BBC 2008). The matter threatened to become a legal case between the BBC, PX, and some of the surveyed mosques (*Independent* 2008), and the report is now unavailable in the think tank's website. The second, *Cities Unlimited* (Leunig and Swaffield 08/2008), claimed the decline of parts of northern England was unlikely to be reversed and that policy incentives should concentrate on the south. Given his perceived closeness to PX and the understandable electoral unpopularity of such an idea, Cameron explicitly distanced himself from the report's findings (*Guardian* 2008a).

Those episodes notwithstanding, by 2008 and after years of sustained growth, PX began to be noticed by competitors and observers of British politics (*Financial Times* 2008b). PX was then seen as a laboratory for cutting-edge Conservative thinking and as the 'government in waiting,' not unlike what had been the case for IPPR in the Labour camp in the 1990s, to which it was often compared (*Guardian* 2008b). The centrality of PX to networks of power, as well as its contribution to the articulation of new centre-right ideas—especially in education and welfare (*Financial Times* 2009)—augured the programme of a future Conservative administration.

However, the political ascent of Conservative modernisers occurred parallel to the 2008 economic crisis. An abrupt economic contraction, bank nationalisations, a credit crunch, a plunge in tax revenues, and rising unemployment and welfare spending set out challenges to financial, fiscal, and monetary policy that few policy actors, let alone think tanks, were in a position to confront. Besides, relationships between more economically focused second wave think tanks and

Cameron's Conservatives had withered somewhat,[41] and while PX had an 'Economics & Social Policy' unit, it tended to favour areas such as education, policing, and welfare (Wade 2013: 167).

Before 2009, PX's economic research was mostly limited to circumscribed policy issues rather than macroeconomics.[42] Proposals by PX's economics unit included reforming welfare-to-work (Bogdanor 03/2004), replacing London buses (Godson 10/2005), and reorganising state pensions (Hillman 03/2008). Partial exceptions to the above are two reports. A first assessed whether the economic growth Britain had enjoyed in the years leading to 2007 was a 'mirage,' sustained by rising debt and a housing boom (Hartwich et al. 11/2007), and a second argued that the tax system should be simplified (Kay et al. 11/2008).

This research agenda was consistent with PX's brand. As modernisers sought to renew Conservative thinking, economic policy ideas that were not limited to fiscal discipline had to be produced. In any case, other think tanks on the right (ASI, IEA) had made the case for free-market policies and a smaller state for decades—even if their links to the Conservatives had 'atrophied' (Wade, op. cit.). Nevertheless, in the context of the largest financial crisis since 1929, in which the main mission for any future government came to be framed as bringing the nation's economy under control (Pirie 2012), relative inattention to financial and fiscal issues "wasn't going to work anymore" (PX interview).[43]

For the above reasons, the crisis had far-reaching effects on PX's research focus, which are conspicuous in their annual accounts Table 6.2. Funds for Economic research almost quadrupled between 2008 and 2009

[41]The perceived distance between David Cameron and new-right think tanks like IEA and CPS was noted at the time by right-of-centre commentators (*Daily Telegraph* 2009).

[42]This need not apply to PX events, which cover more topics than its reports. On September 30th 2008, PX hosted the panel 'Britain after the Credit Crunch' with Philip Hammond MP, then Shadow Chief Secretary to the Treasury, Businesswoman Kim Wiser, and the Business Editor of the *Daily Express* Tracey Boles, accessed 10 March 2016, http://www.policyexchange.org.uk/events/pastevents/item/britain-after-the-credit-crunch?category_id=37.

[43]"[PX] was set up to be a centre-right conservative-minded think tank that does not spend its time talking about tax, fiscal policy, or Europe […] There was space in the market for […] a centre-right think tank that talks about other kinds of things. So in the early years of Policy Exchange they […] focused on things like housing policy, social policy, environmental policy, things of that sort. Now along came the financial crisis, 2007 through 2008, and that wasn't going to work anymore" (PX interview).

Table 6.2 PX research income per unit

Research income (£)	2007	2008	2009	2010	2011	2012	2013
Economics	127,181	169,602	649,240	135,104	319,975	444,108	382,865
Security	158,180	442,039	335,786	353,710	379,458	538,726	231,235
Education	66,100	125,650	179,485	119,400	117,784	101,725	216,850
Crime and justice	27,500	159,850	155,206	116,500	141,825	197,600	188,029
Governance	8250	15,000	20,000	8000			
Health	7800	79,998	34,300	132,328	7500		
Social policy	111,011	139,255	35,000	2000		10,000	7781
Environment		206,600	229,800	152,784	189,651	175,856	153,133
Digital government					3000	174,000	173,580
Wolfson economic prize						586,957	10,000

Source See Table 6.1

(from £169,602 to £649,240), while those for Healthcare, Governance, and Social Policy dwindled.[44] Also linked to their economic work, by the end of our timeframe, PX hosted the Wolfson Economic Prize (see further below), which provided a large stand-alone donation. In line with these numbers, economics became increasingly central for PX, even while the whole of the organisation grew significantly. Between 2007 and 2008, PX produced 12 reports under the label of 'Economics & Social Policy,' of a total of 45 across the organisation, with few focusing on fiscal or financial issues. In 2009–2010, these were 32 out of 94, a significant number of which delved on public deficits and economic instability.

The first major public intervention by PX directly focused on the crisis was the report *Will the splurge work* (McKenzie Smith et al. 11/2008), which assessed whether Labour's taxation plans could balance a recently announced 20 billion pounds of added public borrowing. In it, PX built a case for reducing the deficit rather than for fiscal loosening, arguing that if there were to be a stimulus, tax cuts should be preferred over increased public spending. Later, in April 2009, PX published *What really happened?* (Saltiel and Thomas 02/2009)[45] where they sought to establish the causes of the financial collapse. Their diagnosis was that the crisis was due to global imbalances and low-interest rates, the expansion of the US subprime mortgage market, and, in the UK, a tripartite regulatory arrangement, the government's lack of savings, and the BoE's inflation targeting. Although these reports provide a good indication of PX's broad understanding of the crisis, a more ambitious and focused set of publications would soon follow.

PX's Chief Economist Oliver Hartwich[46] left PX in October 2008, and the organisation was on the lookout for a replacement. To that effect, Andrew Lilico joined in March 2009 (ConservativeHome 2009).

[44]"[PX's rise in economics work] is almost definitely because of the economic crisis. [Before] we didn't feel there was a need for an economics tax and spend kind of unit in PX because the other think tanks did that. But once the economic crisis hit that, [...] was something we had to respond to [...]. So [PX's economic policy unit] went from one person to four or five" (PX interview).

[45]Incidentally, Miles Saltiel has also written for the ASI.

[46]Oliver Hartwich is a member of the Mont-Pèlerin Society and has been employed in several right-leaning think tanks in Australia and New Zealand. According to his own account, the idea of an independent forecaster modelled after the BoE's MPC—which later became the OBR—came in meetings of PX staff with then Shadow Chancellor George Osborne (*Business Spectator* 2010).

Lilico was then Director of Europe Economics, a member of the IEA/*Sunday Times* Shadow Monetary Policy Committee, and had worked at the IFS and the Institute of Directors. He had also authored a CPS report claiming that the de facto nationalisation of banks marked the end of 'private capitalism' (Lilico 2009). Even if some of his ideas were different to those previously espoused by PX—particularly his opposition to the bank bailouts of 2008[47]—under his stewardship the 'Economics & Social Policy' unit would make a strong case for the need for fiscal consolidation. This was the case despite some initial doubts by senior management over whether the magnitude of the proposed cuts could damage PX's moderate reputation. One interviewee said:

> It was felt that the senior levels of PX [...] the directors and so on [...] were sceptical so [researchers] had to go to some effort to persuade them that the cuts [...] weren't so high as to [...] damage their credibility. They were nervous about that". (PX interview)

In June 2009, one of the first PX reports specifically focused on fiscal policy was released, *Controlling public spending: The scale of the challenge* (Atashzai et al. 06/2009). It was authored by Lilico, Adam Atashzai, future Special Advisor to the Prime Minister, and Neil O'Brien, then PX Director. The report argued that a rapid projected increase of public spending as a percentage of GDP (likely to reach 50% by 2010) had to be decisively halted. This was argued on the grounds that: (a) a rising government deficit would reduce growth prospects; (b) preventing an expansion of state expenditure was easier than cutting it later, and; (c) since surveys showed that a majority of the public was against a significant increase in the size of government. Furthermore, of a forecasted £119 billion rise in public debt in 2009, PX estimated that 56% was on consumption rather than capital investment spending (6%) or directly caused by the crisis (38%).

[47]While Saltiel and Thomas (02/2009: 10) claimed banks' bailout "was necessary for the economy – and society – to function," Lilico (2009: 46) pondered in a CPS report: "how bad would the recession have been without the Government's interventions? Could it have been worse than creating a 5%–6% add-on to the recession, spending hundreds of billions of pounds in the process, destroying private capitalism, and forcing the bailing out of other types of company and the enactment of wealth taxes. Was this a better strategy than using the money to cut our taxes or provide other sorts of comfort?"

PX continued with a more extensive report assessing international and historical evidence for financial consolidation (Holmes et al. 11/2009)—which was later cited by renowned economist Carmen Reinhart (Reinhart and Sbrancia, 2011). In this publication, PX argued that fiscal consolidation can promote growth by enabling a looser monetary policy which could "offset the potentially contractionary short-term effects of fiscal tightening" (Holmes et al., op. cit.: 26). This was argued through three economic mechanisms, contra-Keynes: (a) a reduction of public debt, if achieved through fiscal consolidation rather than tax rises, is associated with future economic growth, boosting demand; (b) reduced public spending is linked to a diminishing tax burden, which also encourages demand; (c) fiscal consolidations, "if perceived as permanent and successful," are likely to increase credibility and reduce interest rates on government debt (ibid.: 25–26). Interestingly, in this report, PX implicitly seeks to delimit the scope of what economics argues for (which would leave, for instance, NIESR out):

> [e]ven notorious cases in which it is commonly believed by non-economists that fiscal tightening promoted additional recession turn out to be more complicated than often thought. (Holmes et al. 11/2009: 105)

It is worth remembering that by late 2009 the cleavage between Labour and the Conservatives was, rather than on whether public spending should be curtailed, on the timing and extent of its reduction. This situation was perhaps partly because, as academic research on the media coverage of the financial crisis shows, towards the second half of 2009 the idea that austerity was inevitable had become ubiquitous in the media (Pirie 2012; Berry, 2016). While Labour argued that fiscal consolidation should wait until substantial growth resumed, Conservatives claimed it should begin in earnest as soon as the worst of the recession was over. In that milieu, PX went further by supporting the case for drastic spending reductions, claiming that historical evidence showed that "short sharp pain is better than dragging it out and if you do that you tend to have a better outcome, you tend to have sharper recoveries" (PX interview).

In subsequent reports, PX would continue arguing that public spending cuts would not undermine the recovery, but rather could even boost it (Holmes et al. 04/2010). This argument was based on a literature review of the experience of non-Keynesian effects in successive fiscal consolidations, in which Alesina and Ardagna (2009) feature prominently. Another aspect of their proposals was that to bolster future economic growth, deficit reduction plans should prioritise spending cuts rather than tax hikes in an 80/20 ratio. These ideas were complemented with work on how PX researchers considered the British tax structure should be reformed in a fiscally neutral, growth-friendly way (Lilico and Sameen 03/2010).

Overall, however, it is difficult to assess the precise impact of such reports on the austerity programme implemented by the future Conservative-led government, as they are part of networks of expertise in which many actors, including politicians, journalists, pressure groups, and other think tanks were producing similar arguments. Furthermore, correlation is not causation. A centre-right government could have conceivably arrived independently to a similar plan. And even supposing policymakers were to unreservedly subscribe to a think tank's proposals, the vagaries of politics and public administration would preclude their unqualified implementation. Nevertheless, even if PX's advice were not to be heeded to the letter, their public interventions contributed to producing a political climate where drastic fiscal consolidation came to be seen as reasonable and perhaps inevitable. Interestingly, in parallel to their work on economics, PX investigated how the public perceived government spending cuts through surveys and focus groups (O'Brien 11/2009). This attention to how austerity is viewed by the electorate continued throughout the Coalition government in reports on public attitudes towards fairness, poverty, and welfare reform (O'Brien 04/2011).

More broadly, in the run-up to the 2010 election, PX published *The renewal of government: A manifesto for whoever wins the election* (Clark and O'Brien 03/2010), co-written by PX Director and a leading *Times* columnist. In it, PX posited that, after the election, there should be sweeping reforms to the state across sectors (energy, transport, education), shrinking its size, in consideration of "the need to swap central targets and controls for the right structures and incentives" (ibid.: 12). However, as implicit in the report's title, this was done at arm's length from Conservatives, since:

> [...] before the election all think tanks are, well nearly all, are charities, so there is charity commission requirement, [...] to be politically neutral, to be nonpartisan [...] so although we are centre-right [...] we are not saying we are conservative, a Conservative Party think tank or anything like that. So before the election [...] for about three months we're not allowed to speak publicly or in a very limited way. After the election, we responded to Osborne's emergency budget. (PX interview)

Soon after the 2010 election and the formation of the Conservative-led coalition government, PX began to produce more precise research on where cuts should concentrate, especially after the 2010 October Emergency Budget (Holmes and Lilico 06/2010). PX proposed to cut an average of 25% of discretionary public spending across non-ringfenced departments (i.e., all but Health and International Development; Holmes et al. 11/2010). Through the very exercise of exploring the feasibility of cuts, the 'Economics & Social Policy' unit was put in contact other areas of the organisation, such as education and policing, as well as with civil servants and politicians. In other words, from a general overview of the need for deficit reduction, once Conservatives were in power, PX started assessing where these reductions should take place.[48] One of the most important areas of focus was public service reform, as, according to one interviewee "one of the biggest items of government is staffing. Six million are employed in the public sector. [We were] looking for efficiencies, looking for ways to make the cuts less painful" (PX interview).

48"There is a shift from the general [...] controlling spending, government deficits, [...] to more specific [areas]. And actually later on [...] we started looking at 'well we've done departmental spending, that's already been kind of laid out [...] what is left? And the answer was public sector pay and welfare reform which is obviously a huge chunk which we hadn't done a great deal on, and that became more important in 2012, 2013" (PX interview).

Meanwhile, as the Coalition's spending plans began to gather opposition—such as that by NIESR and NEF—PX sought to defend the austerity programme and see through its implementation. Instances of this support included interventions in the media, as when then Director Neil O'Brien questioned the IFS's claim that spending cuts would trigger a rise in child poverty (*Guardian* 2010), and when he defended austerity and argued that Labour had, throughout the 2000s, drifted away from fiscal responsibility (*Daily Telegraph* 2011). Similarly, Lilico wrote that the austerity programme had to be maintained and that the only policy alternative was even more stringent cuts (ConservativeHome 2011), based on the idea that:

> [w]ith a double dip recession, markets will be more demanding of urgent action, not less, and any sniff that the Coalition is losing the political will to carry its programme through could be disastrous. (Lilico 08/2010)

In that sense, PX's positioning drifted to the right in economics matters, visibly even in areas beyond fiscal consolidation.[49] For instance, in *Beware false prophets* (Saunders 08/2010) a PX external author extensively criticised the statistical arguments behind *The spirit level* (Pickett and Wilkinson 2010), a book that had become central for those arguing against austerity and for the benefits of reducing inequality. After 2010, PX interventions began to concentrate on welfare reform to complement the public spending reduction sought through discretionary department expenditures. Andrew Lilico left PX in 2011, and was replaced by Matthew Oakley, former Economic Advisor to the Treasury. Oakley, as Head of 'Economics & Social Policy' focused much of the unit's research on how to curtail welfare spending.

Two of Oakley's early proposals were that unemployment benefits claimants capable of working should spend more time looking for work (Oakley 05/2011) and for a points-based system for jobseekers'

---

[49]However, one should not exaggerate the organisation's internal coherence, as public disagreements with former members have occurred. In 2012, Alex Morton, then PX Head of Housing, disagreed publicly with Andrew Lilico's assertion that there is no housing shortage in the south of England (Morton 12/2012).

allowance (Doctor and Oakley 11/2011). On the same topic, PX later argued that jobcentres should be assessed on their success in bringing benefits claimants into sustainable full-time employment, and that "[p] art-time employees claiming Universal Credit could be required to provide Jobcentre Plus with evidence to show that they are seeking longer hours, higher pay or higher paid jobs" (Oakley 11/2012: 15). In tandem with this programme, PX investigated public attitudes to workfare and social security through surveys (Holmes 09/2013), finding that a majority had become less supportive of the welfare provision for the long-term unemployed. This is remindful of previous PX research surveying attitudes to fiscal reform, which effectively linked policy proposals to their approval by the electorate—and hence their political desirability.

In parallel, PX began to research on issues pertaining to the financial industry. One of the first publications to that effect was a survey of City workers and senior managers' opinion on London's regulatory regime (Sumpster 12/2010). Later James Barty, a banker and consultant with decades of experience in finance, joined PX as Senior Consultant on Financial Policy. He penned several reports on financial reform, the first being, in the context of the Greek crisis, on the lessons left by Argentina's 2002 sovereign debt default (Barty 03/2012). He later produced public interventions on the need for reforming the BoE (Barty 12/2012), executive compensation (Barty 07/2012), capital requirements (Barty 03/2013), and financial provision for SMEs (Barty 11/2013). In sum, from relatively little work on fiscal, financial, and monetary policy, after 2008 PX had become increasingly active in economic affairs. So much so that, in 2012, PX received over £1m for economic research. That year, the institution of the Wolfson Economics Prize Fund—sponsored by Conservative donor, life peer, and PX trustee Simon Wolfson and the Charles Wolfson Charitable Trust—was inaugurated, which offered £250,000 to the best economic proposal for managing a hypothetical breakup of the European Union.

Ultimately, these transformations pulled PX towards more traditional small-state conservatism, in parallel to the growing primacy of fiscal discipline in Cameron's leadership. In the words of a commentator on the Conservative Party: "[p]ractical financial aspects have certainly hindered the promotion of the 'Big Society' agenda since 2010, and this has meant that aligning fiscal conservatism to the delivery of enhanced social justice has proved to be a conundrum for conservative modernisers" (Williams 2015: 182). Fiscal consolidation brought to the mind of many the image of earlier Conservative administrations, precisely the brand modernisers had once sought to distance themselves from. One interviewee comments:

> The original Cameron Project or the modernising of the Conservative Party was about running public services better [...] So sharing the proceeds of growth idea we are going to match Labour's spending plans [...] That's very much where Policy Exchange was, up to the financial crisis. After that point, the debate on the centre-right changed and we changed [...] It's not hypocrisy [...] it's just the financial crisis happened and [...] you can't share the proceeds of growth if there's no growth. [...] So a lot of [PX] work is about finding economies, finding efficiencies where we possibly can. So [...] maybe we're less, our reputation is less modernised now. (PX interview)

Apropos this transformation, in January 2012, Anthony Seldon—son of Arthur Seldon, IEA's founder—published with PX *The politics of optimism* (01/2012). In it, Seldon argued for a revitalisation of the 'Big Society' agenda to proactively face the crisis, rekindling the moderniser's focus on optimism and social justice within the bounds of fiscal discipline. In that sense, part of the appeal of 'compassionate conservatism' hinged on its capacity to supply a future Tory government with policy ideas in areas such as housing, welfare, and education, with the hope of balancing fiscal tightening with the advancement of social justice (*Guardian* 2012).

One example of this effort to expand Conservative's policy focus pertains housing, where advocacy for increased homeownership became central both to Cameron's government and PX's research agenda. Within this ambit, the report *Making housing affordable* (Morton 08/2010), which proposed increasing the social housing supply through a 'community-controlled' planning system, was awarded *Prospect* 'think tank publication of the year' (Prospect 2010). A subsequent PX publication on housing by Morton (08/2012) would be endorsed by the then Housing Minister, Prime Minister, and Chancellor.

Two further areas in which PX attempted to supersede the tension between modernisation and austerity—ambitious policy innovations under budgetary constraints—include digital government and behavioural economics. On the former, a PX unit was established in 2011 which, with the support of technology companies such as Microsoft, EE, and Intuit, explored ways in which government could be made more efficient and public data more transparent (Yiu 03/2012; Fink and Yiu 09/2013). On the latter PX published several reports across its units, drawing from behavioural economics to achieve better policy outcomes, for instance on measuring inflation (Shiller 05/2009) and environmental policy (Newey 01/2013). In July 2011, PX hosted an event with Richard Thaler, co-author of the influential book *Nudge* (Sunstein and Thaler 2008) and advisor to the government's newly inaugurated Behavioural Insights Team.[50]

Concerning PX's precise influence, although causation is difficult to establish between their ideas and the policies of the Coalition government, the correlation is sometimes staggering. To present its own imprint, PX produced a diagram tracing their previous reports and how they claim they influenced subsequent government measures (PX 02/2012)—showcasing, once again, that PX's very brand hinges on impact. Among the examples of influence that diagram covers, perhaps the clearest concern education and policing. On the former one can mention their proposal for a 'pupil' or 'advantage' premium (Leslie and O'Shaughnessy 12/2005)—state funding on a per capita basis for students from disadvantaged backgrounds—which became policy in 2011. Also noteworthy is the establishment of free schools (independent schools governed by non-profit trusts), an idea advanced in early PX publications (Hockley and Nieto 05/2005), which became law under the 2010 Academies Act. On policing, PX first report (Loveday and Reid 01/2003)—as well as later ones (Chambers 11/2009)—argued for greater local participation in the governance of the police through the democratic election of commissioners, a measure that is part of the 2011 Police Reform and Social Responsibility Act.

[50]Accessed 15 March 2016, http://www.policyexchange.org.uk/modevents/item/turning-behavioural-insights-into-policy-with-richard-thaler-author-of-nudge-and-advisor-to-number-10.

Besides those instances, perhaps PX's wider political import is also apparent in the events they held and the trajectory of their 'alumni.' Throughout the timeframe of this research, PX's events programme became an ever more important aspect of the organisation, strengthening their position as convenors of political elites and offering a space for the discussion and coordination of policy ideas, especially among Conservatives MPs and ministers.[51] Events reached a frequency of almost one per week (PX 02/2012) and became an important source of revenue. In addition, given the renown of many of its guests, this think tank's press coverage was propelled by the sheer number of set speeches by senior politicians it hosted (see Chapter 7).

With regards to PX staff turnover and their trajectories, although functioning as a party 'revolving door' could conceivably present challenges for the retention of talent, it certainly added to the organisation's reputation as a node for the articulation of the political elite. Indeed, the number of former PX members who became (to name but one title) special advisers to government departments between 2007 and 2013 is impressive (e.g., Amy Fisher, Alex Morton, Neil O'Brien, Ruth Porter). In addition, throughout the Cameron years PX alumni featured consistently across government consultations, committees, and commissions. For instance, after PX Neil O'Brien became Deputy Chair at the Social Mobility and Child Poverty Commission, and Matthew Oakley was appointed to the DWP's Social Security Advisory Committee (Gov.uk 2012d, 2013d). Two further examples of PX's centrality are worthy of note to show how this think tank can both 'send' and 'receive' important figures of the British political right. One is a former employee who

[51]By way of illustration, senior politicians who have spoken at PX events between 2007 and 2013 include then Home Secretary and future Prime Minister Theresa May, as well as then Immigration Minister Damian Green, on immigration reform (Gov.uk 2010a, 2012a); Nick Herbert on criminal justice reform (Gov.uk 2010b); Michael Gove on pension reform (Gov.uk 2011); Lord Freud, Minister for Welfare Reform, on employment outcomes (Gov.uk 2013a); then Exchequer Secretary to the Treasury David Gauke MP on tax avoidance (Gov.uk 2012b); Michael Fallon on postal service reform (Gov.uk 2013b); David Lidington MP on the European single market (Gov.uk 2010c); Francis Maude on civil service reform (Gov.uk 2013c); David Willetts on growth policies and high tech industrial strategy (Gov.uk 2012c) Gregg Clark on the 'Big Society' (Gov.uk 2010d); and General Richards on national security strategy (Gov.uk 2010e).

became a central political figure and the other is an influential policy expert who later joined PX; respectively James O'Shaughnessy and Steve Hilton. The first, after leaving PX management in 2007, co-authored the 2010 Conservative general election manifesto and became David Cameron's Director of Policy. The second was Director of Strategy for the Prime Minister and became PX Visiting Scholar in 2015. These days Hilton has his own show at Fox News.

The Politics of the Reasonable

I stated at the beginning of this book that I would not focus on the policy impact of think tanks, but on their intellectual and institutional transformations following the crisis. However, in PX's case, these two issues are difficult to disentangle. As the think tank most closely aligned with Tory modernisers, by 2008 the fate of PX and of Cameron's Conservatives had become ever more entwined, and there was little perceived distance between its ideas and those of the party leadership—which explains the pressures Cameron faced to publicly disavow the recommendations of PX report *Cities Unlimited* (Leunig and Swaffield 08/2008). By then PX had become, in the eyes of both supporters and critics, Cameron's 'policy shop.' As such, in the run-up to the 2010 election, PX began to devise proposals that could plausibly be considered a proxy for the policies likely to be implemented by a future Tory government. It can be said that, instead of moving the 'Overton window' by pressuring politicians from a distance, as 1970s second wave think tanks such as the ASI had sought to, PX's role was rather to determine where its centre is. Unsurprisingly, this vocation for centrality generated tensions between their reputation for both political impact and intellectual independence. In the work of Richard Cockett, among the most prominent chroniclers of the rise of the 1970s new-right:

> The IEA and the CPS were set up by a bunch of mavericks. They really were trying to change the world. They were at war with the Conservative Party hierarchy. It's very different with Policy Exchange. These guys are all quite young. They want to make a career in politics. They move smoothly into the party. (Cockett, in *Guardian* 2008b)

PX's centrality also meant that it had to keep apace with the vicissitudes of politics. Like Cameron, they supported both the 'Big Society' programme and, after 2009–2010, austerity. Producing public interventions in support of both these policy agendas required some change of tack and image, which made clear that a vocation for political influence demands a degree of intellectual pliability. In this sense, although the *hysteresis* hypothesis as set out in Chapter II correctly predicted that PX would aspire to form a plan for government and see through its implementation, it risked rendering invisible the fact that the plan they ultimately defended could itself involve some sort of change. That perhaps implies that the position PX sought to defend after the crisis was not first and foremost intellectual, but political. In other words, in the years since Cameron became leader PX had succeeded in becoming 'Prince's courtiers,' with all the pomp and regalia that entails, but in 2010 the Prince's whims had changed. The circumstances after 2008—some would argue for economic, others for political reasons—did not allow for straightforward advocacy of 'Compassionate Conservatism' as envisaged in PX's early days. Their public interventions, especially in the 'Economics & Social Policy' unit, could not be the same they set out before the crisis. In tandem, across other parts of the organisation, the need to cut public spending became a precondition for setting forth any new proposal.

Overall, a constant in PX's public interventions was that they sought to position the think tank as a producer of innovative yet politically reasonable and implementable centre-right ideas. In order to occupy that space, attentiveness to their political environment was paramount, perhaps to a greater degree than for think tanks such as NEF, which continued to produce research into *The great transition* long after it became politically untimely. For that reason, as perceived problems and 'windows of opportunity' (Kingdon 2003) shifted, so did PX's research agenda. As it became apparent Healthcare was not an area were comprehensive reforms would take place and that NHS spending would be ringfenced, PX's Health unit withered.

In relation to economics, PX's role was, more than to convince the Conservative leadership of the need for drastic public spending cuts through the public conversation (as the ASI or the IEA would), to propose policies that could implement these cuts in a practical and, crucially, politically acceptable fashion (Pautz 2016). According to research on the media coverage of the crisis (Pirie 2012; Berry 2016) by the time PX had released their series of reports on the desirability of fiscal consolidation, the idea that government profligacy was the most urgent aspect of the crisis had become pervasive. In that juncture, PX was in a privileged position—less accessible to the ASI, for instance—to propose an austerity programme that could be perceived as measured and close to the political mainstream, rather than be dismissed as strictly ideological. Still, PX's precise influence on Osborne's plans is not clear, and there were discrepancies between interviewees. One said:

> Victory has a thousand fathers, defeat is an orphan [...] so I wouldn't claim [...] for example that fiscal consolidation wasn't going to happen anyway. I think we were read by some senior people, and I think maybe we helped frame [it], but it's not, it's not decisive. (PX interview)

While another claimed:

> It definitely wasn't that the Conservatives or the LibDems or anybody was telling PX what to think, PX was telling them what to think [...] the integration was in that direction [...] Conservative ideas and PX ideas were almost the same thing. But that was because the Conservatives wanted all the things that PX did. (PX interview)

Even if the above quotes differ on the direct impact they believe PX's work had, they are compatible at a more abstract level—whatever the direction of the relationship. One way of thinking about PX's role is through Ladi's (2011) work on think tanks in policy shifts. She posits that the purpose of ideas and research in policy is often symbolic rather than instrumental: "discourse is central in the preparation of policy shifts, and think tanks are key carriers of both coordinative and communicative discourse" (Ladi, op. cit.: 217). That would imply that, given the close relationship between PX and the Conservative leadership, their move towards supporting stringent public spending cuts was an indication that the consensus in the centre-right was shifting. What was needed at such a juncture was an academically defensible policy discourse that could make the case that austerity was necessary, workable, and politically defensible. In such a way, PX reports on the deficit sought to reconcile respectable economic expertise with politically felicitous policies.

Another index of PX's role as 'mediators' and 'moderators' of research, policy, and politics concerns its remarkable sensitivity to polling and to what government can achieve. This attentiveness is another indication that PX's aim is mainly to produce policy proposals that are compatible with a given politics, working within the bounds of a 'political' positioning that necessitates always being seen as reasonable and moderate. It is not casual that PX authors define their organisation much more frequently as 'centre-right' than as 'free market'—in contrast, for instance, to the ASI.

However, even within the fuzzy bounds of the right flank of the political 'centre,' the emergence of austerity begot an unavoidable contradiction between the discourse of political modernisation and that of the need for 'belt-tightening'—especially considering the weight Thatcher still has on British right-of-centre politics. In that predicament, much of PX research attempted to render 'Big Society' thinking and austerity compatible, for instance, by generating efficiencies through the work of the 'Digital Government' unit. However, attempts to ease out this tension were difficult to achieve in all instances and across all policy areas, especially given that the think tank remained closely associated with Cameron's government. In the words of an interviewee "PX's thing of being a sort of centrist to right think tank […] became much less relevant" (PX interview). As public discourse and the Conservative party lurched to the right, so did PX.

In summary, much of what PX's public interventions sought was to provide policies that were both 'politically reasonable' and 'technically feasible.' Given this think tank's privileged position as the crucible of centre-right policy thinking, it could moderate the contact between high-level official politics and practical economic policymaking. Within this ambit, however, much could be under discussion, and internal coherence was not always paramount. In that vein, perhaps owing to the weight of the New Labour years—shown in an orientation towards the political centre and in the adherence to a 'what works' policy discourse—it was necessary to present a measure of flexibility that allowed for a mercurial corporate view. Here it is worth noting what Anthony Seldon wrote in a 2012 PX publication that the Big Society programme "lacked the intellectual coherence enjoyed by the domestic agendas of the three successful [Labour] governments: in critical respects, deep thinking on the Big Society had not taken place" (Seldon 01/2012: 3).

An interesting juncture in which such flexibility and lack of coherence is visible concerns PX's research agenda with regards to immigration. The organisation had little to no research on the matter before 2014, well after it had become a significant political and policy issue. That changed, perhaps because immigration became an area where reforms were bound to happen and since the need for immigration controls had become 'commonsensical' on the right, even among modernisers. In 2015, David Goodhart—formerly head of Demos and who, incidentally, had clashed repeatedly with NIESR's Jonathan Portes (see Chapter 5)—joined PX as head of its new 'Demography, Immigration and Integration' unit. In 2017, this unit released a report entitled *Racial-self interested is not racism* (Kaufmann 05/2017), which would have been unthinkable in 2010. This is but one example of PX's sensitivity to what is thought to be 'reasonable' in their political sector at any given time. One could also add their later proposal for 'Irexit,' for Ireland to join the UK in leaving the EU (Bassett 06/2017)—which makes, in 2019, for an interesting read.

In that sense, PX's orientation towards pragmatism and moderation is dependent on the fact that what is perceived as mainstream and moderate changes over time, which tints what kind of expertise in economics or immigration is considered relevant at any given juncture. It suffices to think of the diametrically different views PX and NIESR authors have on what the economic consensus is to illustrate that point. In that sense, occupying a privileged conduit between fields, and unlike NIESR, NEF, or even the ASI—whose proposals would seem at least at times to go against the tide of opinion on the right (e.g., on immigration)—PX's role would be instead to channel the tide; to mediate between incompatible demands from across fields, if ultimately driven by their vocation for political relevance. To borrow a concept from the Neoinstitutionalist literature, PX's mission would be to produce "bounded innovation" (Weir 1992: 189).

In the following and last chapter, I compare the four case-studies and set out a research programme on the wider intellectual and political role of think tanks, based on the concept of 'moderation.' I argue that politically influential policy organisations such as PX can be central conduits for the moderation of demands from diverse publics or, in other words, for determining what should be considered moderate and reasonable in spaces that are not solely determined by evidence and 'pure' expertise.

References

Alesina, A., & Ardagna, S. (2009). *Large changes in fiscal policy: Taxes versus spending* (NBER working papers, 15438). Accessed 10 April 2015. http://www.nber.org/papers/w15438.pdf.

BBC. (2008). *Policy Exchange dispute—Update*. Accessed 15 February 2016. http://www.bbc.co.uk/blogs/newsnight/2008/05/policy_exchange_dispute_update.html.

Berry, M. (2016). No alternative to austerity: How BBC broadcast news reported the deficit debate. *Media, Culture and Society, 38*(6), 844–863.

Business Spectator. (2010). *Oliver Marc Hartwich: It's time Henry had an umpire*. Accessed 3 April 2016. https://web.archive.org/web/20120415235526/http://www.businessspectator.com.au/bs.nsf/Article/MRRT-RSPR-Treasury-statistics-macroeconomics-pd20100720-7J7PX?OpenDocument.

ConservativeHome. (2009). *Policy Exchange appoints Andrew Lilico as its Chief Economist*. Accessed 12 March 2016. http://conservativehome.blogs.com/torydiary/think_tanks_and_campaigners/.

ConservativeHome. (2010). *Philip Booth: Why I have reservations about the "Nudge" philosophy*. Accessed 12 March 2016. http://www.conservativehome.com/platform/2010/03/philip-booth-why-i-have-reservations-about-the-nudge-philosophy.html.

ConservativeHome. (2011). *Andrew Lilico: Stick to the course, Mr Osborne. Don't let Ed Balls bore you into submission*. Accessed 21 April 2016. http://www.conservativehome.com/platform/2011/06/yaaawwn.html.

Daily Telegraph. (2001). *Norman still selling Portillo's dream*. Accessed 15 March 2016. http://www.telegraph.co.uk/news/uknews/1334767/Norman-still-selling-Portillos-dream.html.

Daily Telegraph. (2009). *If David Cameron wants to govern, he should stop being afraid of ideas*. Accessed 24 April 2016. http://www.telegraph.co.uk/comment/columnists/simonheffer/5700830/If-David-Cameron-wants-to-govern-he-should-stop-being-afraid-of-ideas.html.

Daily Telegraph. (2011). *Neil O'Brien: Why the deficit deniers are deliberately missing the point*. Accessed 15 March 2016. https://web.archive.org/web/20160326235251/http://blogs.telegraph.co.uk/news/neilobrien1/100074386/why-the-deficit-deniers-are-deliberately-missing-the-point/.

Denham, A., & Garnett, M. (1998). *British think tanks and the climate of opinion*. London: UCL Press.

Denham, A., & Garnett, M. (2006). What works? British think tanks and the end of ideology. *Political Quarterly, 77*(2), 156–165.

Evans, K. (2011). 'Big society' in the UK: A policy review. *Children and Society, 25*, 164–171.

Financial Times. (2008a). *Think tank feels pinch as rival cashes in*. Accessed 27 February 2016. http://www.ft.com/cms/s/0/a8990ad0-7453-11dd-bc91-0000779fd18c.html.

Financial Times. (2008b). *Policy Exchange powers party's 'liberal revolution.'* Accessed 2 March 2016. http://www.ft.com/cms/s/0/3006a6c4-26d1-11dd-9c95-000077b07658.html.

Financial Times. (2009b). *Meet the new Tory establishment*. Accessed 5 March 2016. http://www.ft.com/cms/s/2/ac5f0298-af38-11de-ba1c-00144feabdc0.html.

Gov.uk. (2010a). *Immigration: Home Secretary's speech of 5 November 2010*. Accessed 20 February 2016. https://www.gov.uk/government/speeches/immigration-home-secretarys-speech-of-5-november-2010.

Gov.uk. (2010b). *Nick Herbert's speech to the Policy Exchange*. Accessed 25 February 2016. https://www.gov.uk/government/speeches/nick-herberts-speech-to-the-policy-exchange.

Gov.uk. (2010c). *David Lidington: The Single Market is "essential to this government's agenda for trade and competitiveness."* Accessed 1 March 2016. https://www.gov.uk/government/news/the-single-market-is-essential-to-this-government-s-agenda-for-trade-and-competitiveness.

Gov.uk. (2010d). *Greg Clark: Three actions needed to help the Big Society grow*. Accessed 3 March 2016. https://www.gov.uk/government/news/three-actions-needed-to-help-the-big-society-grow.

Gov.uk. (2010e). *General Richards: Trading the perfect for the acceptable*. Accessed 3 March 2016. https://www.gov.uk/government/news/cds-on-the-sdsr-trading-the-perfect-for-the-acceptable.

Gov.uk. (2011). *Michael Gove on public sector pension reforms*. Accessed 3 March 2016. https://www.gov.uk/government/speeches/michael-gove-on-public-sector-pension-reforms.

Gov.uk. (2012a). *Damian Green: Making immigration work for Britain*. Accessed 28 February 2016. https://www.gov.uk/government/speeches/damian-greens-speech-on-making-immigration-work-for-britain.

Gov.uk. (2012b). *David Gauke: Where next for tackling tax avoidance?* Accessed 20 February 2016. https://www.gov.uk/government/speeches/speech-by-exchequer-secretary-to-the-treasury-david-gauke-mp-where-next-for-tackling-tax-avoidance.

Gov.uk. (2012c). *David Willetts: Our hi-tech future*. Accessed 3 March 2016. http://webarchive.nationalarchives.gov.uk/20121212135622/http://www.bis.gov.uk//news/speeches/david-willetts-policy-exchange-britain-best-place-science-2012.

Gov.uk. (2012d). *Alan Milburn and Neil O'Brien set to lead the drive to improve social mobility and reduce child poverty*. Accessed 25 April 2016. https://www.gov.uk/government/news/alan-milburn-and-neil-obrien-set-to-lead-the-drive-to-improve-social-mobility-and-reduce-child-poverty.

Gov.uk. (2013a). *Lord Freud: Improving employment outcomes*. Accessed 20 February 2016. https://www.gov.uk/government/speeches/improving-employment-outcomes–2.

Gov.uk. (2013b). *Michael Fallon: Royal mail—Ensuring long term success*. Accessed 1 March 2016. https://www.gov.uk/government/speeches/royal-mail-ensuring-long-term-success.

Gov.uk. (2013c). *Francis Maude: Ministers and mandarins—Speaking truth unto power*. Accessed 1 March 2016. https://www.gov.uk/government/speeches/ministers-and-mandarins-speaking-truth-unto-power.

Gov.uk. (2013d). *Government announces new member to the Social Security Advisory Committee*. Accessed 25 April 2016. https://www.gov.uk/government/news/government-announces-new-member-to-the-social-security-advisory-committee.

Guardian. (2005). *Full text: David Cameron's speech to Policy Exchange*. Accessed 15 March 2016. http://www.theguardian.com/politics/2005/jun/29/speeches.conservatives.

Guardian. (2008a). *Cameron condemns 'insane' report on northern cities*. Accessed 15 March 2016. http://www.theguardian.com/politics/2008/aug/13/davidcameron.conservatives1.

Guardian. (2008b). *What can they be thinking?* Accessed 28 February 2016. http://www.theguardian.com/politics/2008/sep/26/thinktanks.conservatives.

Guardian. (2010). *Spending cuts 'will see rise in absolute child poverty'*. Accessed 28 March 2016. http://www.theguardian.com/politics/2010/dec/16/spending-cuts-rise-absolute-child-poverty.

Guardian. (2012). *Compassionate conservatives find it's time to think again*. Accessed 28 March 2016. http://www.theguardian.com/politics/2012/mar/05/compassionate-conservatives.

Hayton, R. (2012). *Reconstructing conservatism? The Conservative Party in opposition, 1997–2010*. Manchester: Manchester University Press.

Independent. (2008). *Tories' favourite think tank sued by Muslim group*. Accessed 18 May 2016. http://www.independent.co.uk/news/uk/politics/tories-favourite-think-tank-sued-by-muslim-group-897548.html.

Kingdon, J. (2003). *Agendas, alternatives and public policies*. New York: Longman.

Ladi, S. (2011). Think tanks, discursive institutionalism and policy change. In G. Papanagnou (Ed.), *Social science and policy challenges: Democracy, values and capacities*. Paris: UNESCO.

Lilico, A. (2009). *What killed capitalism? The crisis: What caused it and how to respond*. London: Centre for Policy Studies.

McLennan, G. (2004). Travelling with vehicular ideas: The case of the third way. *Economy & Society, 33*(4), 484–499.

Norman, J. (2010). *The big society: The anatomy of the new politics*. Buckingham: University of Buckingham Press.

Pautz, H. (2012). The think tanks behind 'cameronism'. *British Journal of Politics and International Relations, 15*(3), 362–377.

Pautz, H. (2016). Managing the crisis? Think tanks and the British response to global financial crisis and great recession. *Critical Policy Studies, 11*(2), 191–210 [Online early access].

Pickett, R., & Wilkinson, K. (2010). *The spirit level: Why more equal societies almost always do better*. London: Allen Lane.

Pirie, I. (2012). Representations of economic crisis in contemporary Britain. *British Politics, 7*(4), 341–364.

Prospect. (2010). *Think tank of the year awards—The winners*. Accessed 20 February 2016. http://www.prospectmagazine.co.uk/politics/think-tank-of-the-year-awards-the-winners.

Reinhart, C., & Sbrancia, M. B. (2011). *The liquidation of government debt* (NBER working papers, 16893). Accessed 12 March 2016. http://www.nber.org/papers/w16893.pdf.

Schlesinger, P. (2009). Creativity and the experts: New Labour, think tanks, and the policy process. *International Journal of Press/Politics, 14*(1), 3–20.

Stone, D. (2007). Recycling bins, garbage cans or think tanks? Three myths regarding policy analysis institutes. *Public Administration, 85*(2), 259–278.
Sunstein, C., & Thaler, R. (2008). *Nudge: Improving decisions about health, wealth, and happiness*. New Haven: Yale University Press.
Transparify. (2016). *How transparent are think tanks about who funds them 2016? A survey of 200 think tanks in 47 countries worldwide*. Accessed 20 March 2016. http://static1.squarespace.com/static/52e1f399e-4b06a94c0cdaa41/t/5773022de6f2e1ecf70b26d1/1467154992324/Transparify+2016+Think+Tanks+Report.pdf.
Wade, R. (2013). *Conservative Party economic policy: From Heath in opposition to Cameron in coalition*. Basingstoke: Palgrave Macmillan.
Weir, M. (1992). Ideas and the politics of bounded innovation. In S. Steinmo, K. Thelen, & F. Longstreth (Eds.), *Historical institutionalism in comparative analysis*. Cambridge: Cambridge University Press.
Williams, B. (2015). *The evolution of Conservative Party social policy*. London: Palgrave Macmillan.

Think Tank Reports and Blog Posts (PX, available at Policyexchange.org.uk)

Atashzai, A. Lilico, A., & O'Brien, N. (06/2009). *Controlling public spending: The scale of the challenge.*
Barty, J. (03/2012). *Sovereign default: Lessons for Europe from Argentina's default.*
Barty, J. (07/2012). *Executive compensation: Rewards for success not failure.*
Barty, J. (12/2012). *Reform of the Bank of England: A new bank for a new governor.*
Barty, J. (03/2013). *Capital requirements: Gold plate or lead weight?*
Barty, J. (11/2013) *Financing small and growing firms.*
Bassett, R. (06/2017). *After Brexit, will Ireland be next to exit?*
Bogdanor, A. (03/2004). *Not working: Why workfare should replace the New Deal.*
Bradley, K., & Hakim, C. (01/2008). *Little Britons: Financing childcare choice.*
Chambers, M. (11/2009). *Partners in crime: Democratic accountability and the future of local policing.*
Clark, R., & O'Brien, N. (03/2010). *The renewal of government: A manifesto for whoever wins the election.*
Davies, C., & Lim, C. (03/2008). *Helping schools succeed: Lessons from abroad.*

Davies, R., & Mitchell, L. (11/2008). *Building Bridges: Philanthropy strengthening communities.*
Doctor, G., & Oakley, M. (11/2011). Something for nothing: Reinstating conditionality for jobseekers.
Drayson, K. (11/2013). *Park land: How open data can improve our urban green spaces.*
Evans, A., & Hartwich, O. (06/2005). *Unaffordable housing: Fables and myths.*
Evans, A., & Hartwich, O. (01/2007). *The best laid plans: How planning prevents economic growth.*
Featherstone, H., & Whitham, L. (07/2010). *Careless: Funding long-term care for the elderly.*
Fink, S., & Yiu, C. (09/2013). *Smaller, better, faster, stronger: Remaking government for the digital age.*
Flint, R., & Skelton, D. (07/2012). *What's in a name? Is there a case for equal marriage?*
Ganesh, J., & Norman, J. (06/2006). *Compassionate conservatism: What is it—why we need it.*
Geoghegan, R. (09/2012). *Future of corrections: Exploring the use of electronic monitoring.*
Godson, D. (10/2005). *Replacing the routemaster: How to undo Ken Livingstone's destruction of London's best ever bus.*
Hartwich, O., Lipson, B., & Schmieding, H. (11/2007). *More mirage than miracle: Assessing the UK's economic performance.*
Helm, D. (12/2008). *Credible energy policy: Meeting the challenges of security of supply and climate change.*
Hillman, N. (03/2008). *Quelling the pensions storm: Lessons from the past.*
Hockley, T., & Nieto, D. (05/2005). *Hands up for school choice! Lessons from school choice schemes at home and abroad.*
Holmes, E. (09/2013). *Work fair?*
Holmes, E., & Lilico, A. (06/2010). *Controlling public spending: Pay, staffing and conditions in the public sector.*
Holmes, E., Lilico, A., & Sameen, H. (11/2009). *Controlling spending and Government deficits: Lessons from history and international experience.*
Holmes, E., Lilico, A., & Sameen, H. (04/2010). *The cost of inaction: Why cutting spending will boost recovery, even in the short term.*
Holmes, E., Lilico, A., & Sameen, H. (11/2010). *Controlling public spending: How to cut 25%.*
Jenkins, S. (11/2004). *Big bang localism: A rescue plan for British democracy.*

Kaufmann, E. (05/2017). *'Racial self-interest' is not racism: Ethno-demographic interests and the immigration debate.*
Kay, L., Martin, D., & Smith, N. (11/2008). *The cost of complexity: How Britain's tax system strangles the economy and reduces British competitiveness.*
Leslie, C., & O'Shaughnessy, J. (12/2005). *More good school places.*
Leunig, T., & Swaffield, J. (08/2008). *Cities unlimited: Making urban regeneration work.*
Lilico, A. (08/2010). *Five politically relevant things about where we are on the economy.*
Lilico, A. (02/2011). *On fairness.*
Lilico, A., & Sameen, H. (03/2010). *Taxation, growth and employment.*
Loveday, B., & Reid, A. (01/2003). *Going local: Who should run Britain's police?*
MacEoin, D. (10/2007; unavailable). *The hijacking of British Islam.*
Maude, F. (03/2012). *Ten years of modernisation: Looking back and the challenges ahead.*
McKenzie Smith, J., O'Brien, N., & Thomas, H. (11/2008). *Will the splurge work?*
Mirza, M., Senthilkumaran, A., & Ja'far, A. (01/2007). *Living apart together: British Muslims and the paradox of multiculturalism.*
Morton, A. (08/2010). *Making housing affordable: A new vision for housing policy.*
Morton, A. (08/2012). *Ending expensive social tenancies: Fairness, higher growth and more homes.*
Morton, A. (12/2012). *Crisis? What crisis? Why Andrew Lilico is wrong to say the UK has no shortage of housing.*
Newey, G. (07/2011). *Boosting energy IQ: UK energy efficiency policy for the workplace.*
Newey, G. (01/2013). *Smarter, greener, cheaper: Joining up domestic energy efficiency policy.*
Norman, J. (12/2008). *Compassionate economics.*
Oakley, M. (05/2011). *No rights without responsibility: Rebalancing the welfare state.*
Oakley, M. (11/2012) *Welfare reform 2.0: Long-term solutions, not short-term savings.*
O'Brien, N. (11/2009). *Tax and spending: Views of the British public.*
O'Brien, N. (04/2011). *Just deserts? Attitudes to fairness, poverty and welfare reform.*
PX. (02/2012). *Ten years of shaping the policy agenda.*

Reid, A. (03/2005). *Taming terrorism: It's been done before.*
Saltiel, M., & Thomas, H. (02/2009). *What really happened? Understanding the credit crunch.*
Saunders, P. (08/2010). *Beware false prophets: Equality, the good society and The Spirit Level.*
Schiller, R. (05/2009). *The case for a basket: A new way of showing the true value of money.*
Seldon, A. (01/2012). *The politics of optimism.*
Sumpster, T. (12/2010). *Not with a bang but a whimper: Are we undermining the future of financial services in Britain?*
Tinsley, M. (06/2012). *Too much to lose understanding and supporting Britain's older workers.*
Yiu, C. (03/2012). *A right to data: Fulfilling the promise of open public data in the UK.*

7
Conclusions: Intervening on Shifting Sands

This last chapter is dedicated to three objectives. The first is to assess to what extent think tanks behaved in a predictable manner, as set out in the *hysteresis* model of Chapter 2. The second is to compare how the public interventions of these think tanks occurred in practice and what changes they reveal, both of think tanks' own functioning and of their environment. The third is to propose understanding the changes think tanks underwent—and their place in politics and policy—through their role as 'moderators' situated between unstable fields.

Testing the *Hysteresis* Hypothesis

Earlier in this book, I posited that the most probable scenario after the crisis was that think tanks would not substantially change their core views and that an informed observer could broadly anticipate what they would say. Given that the efforts required to establish a reputation and a brand are considerable, I surmised that a radical shift in their public interventions was improbable. In Bourdieusian terms, this means a *hysteresis* effect would hold sway. Following this line of thinking, come

M. González Hernando, *British Think Tanks After the 2008 Global Financial Crisis*, Palgrave Studies in Science, Knowledge and Policy, https://doi.org/10.1007/978-3-030-20370-2_7

2008 one could expect a left-of-centre think tank to blame the deregulation of finance, and a sense of disappointment would follow after public spending cuts became official policy. Conversely, a free-market think tank would most probably blame public officials and regulation for the debacle in lieu of financial speculators, greed, or the free market itself. An academic-technocratic organisation would most likely try to present itself as a neutral arbiter and argue for the supremacy of economic evidence, and one supporting a party set to become government would devise a plan and defend it through its implementation.

This null hypothesis proved mostly correct, or at least enough so that examples to confirm it were not difficult to find. Nevertheless, the purpose of this model is not to corroborate my suspicions, but to detect where they were inadequate. After going over the four case-studies, I believe this conjecture is insufficient on at least two accounts. First, as its focus is on the content of public interventions, it risks overlooking significant changes in think tanks' modes of public engagement. Second, this model says little of the mechanisms and levels in which intellectual change may happen, overestimating think tanks' internal coherence.

Concerning the first point, as the model I put forward focused mostly on the substantive, cognitive content of think tanks' work, it did not account for the medium and fashion in which content was conveyed. Often, the most interesting transformations concerned the format of think tanks' public interventions rather than their argument. A case in point is NIESR, which had spent decades producing econometric research for specialist audiences rather than for mass media, let alone social media. After the crisis, NIESR saw a sharp rise in the coverage of their work, propelled by growing public attention on hitherto niche macroeconomic debates. However, this came with little perceived influence, as they saw the government pursue a policy agenda they considered to be cavalier with evidence. As a response, a new director in 2011 incentivised greater engagement with non-specialists. NIESR's presence in social media and op-ed pages, negligible in 2009, became ever more noticeable. These efforts, to be sure, were not necessarily linked to shifts in the substantive content of their work, though they did occur in tandem with a more outspoken stance towards official policy. Succinctly put, while the economics behind NIESR's publications may have remained relatively stable, their mode of public engagement changed considerably. This belied a shift from targeting a narrow academic and policy elite to seeking to influence a larger audience, which occurred in parallel to crucial institutional changes, especially in relation to their funding sources.

Developments such as these can be linked to shifts in the environment in which think tanks operate, especially in relation to the publics its work intends to reach. To expand on this point, it is useful to examine their media coverage. After all, the format of a think tank's public interventions is linked to what type of audience it seeks to inform, and traditional media outlets are a privileged conduit for reaching some of these audiences. The graph (Fig. 7.1) shows think tanks' yearly mentions across Britain's four most circulated daily broadsheet. From these numbers, it is readily evident that their media presence expanded considerably after 2008. Total mentions for all cases went from 374 in 2007, and 481 in 2008, to 513 in 2009, and 604 in 2010. To be sure, this increase could happen without direct planning or action by think tanks, as the case the growing interest on NIESR's forecasts shows. However, for institutions whose primary policy interventions are in the form of commentary (particularly the ASI) or policy reports (e.g., NEF), this growth may be linked to an expansion of their output (see Fig. 2.1, Chapter 2). Meanwhile, the media coverage of well-connected policy institutes can be enhanced by their hosting of speeches and announcements of senior politicians—e.g., the case of PX, especially after 2010.

These numbers also provide evidence that the presence of each organisation varies across broadsheets—which can be associated to their political proclivities. For instance, while NEF enjoys considerable visibility in the left-of-centre *The Guardian*, it lags behind in right-leaning papers such as the *Daily Telegraph* and *The Times*. There are also significant variations in the capacity in which each think tank is presented. For example, while most of ASI's coverage is in the form of commentary, much of NIESR's corresponds to citations of their numbers, especially in the *Financial Times*. In this sense, the larger volume of opinion pieces written by NIESR staff as years went by is an index of important shifts

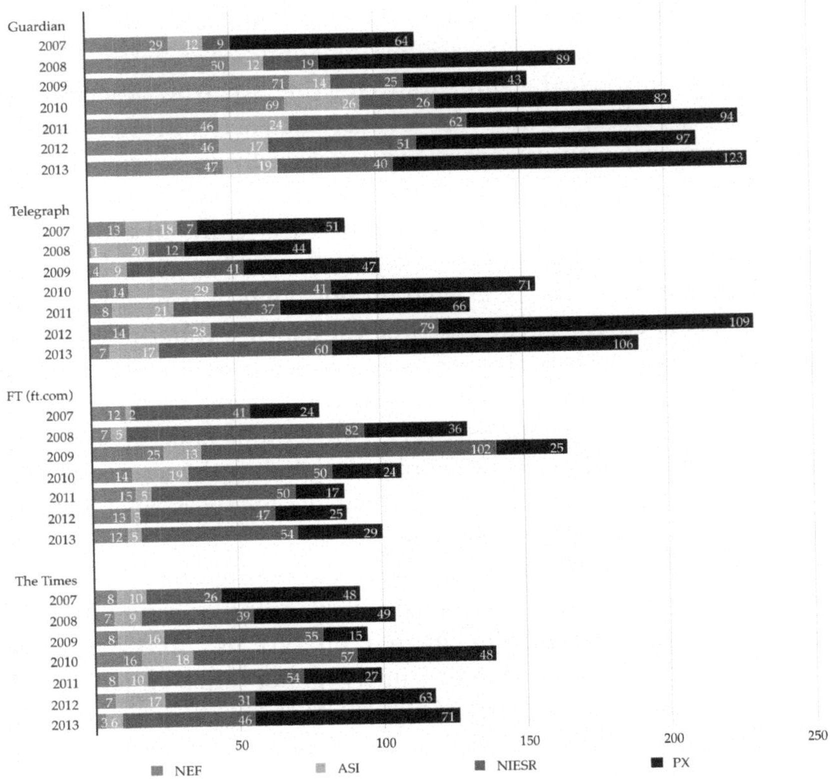

Fig. 7.1 Think tanks' coverage in broadsheets (per year, per source)

in both intellectual (how management understood their mission) and institutional terms (the support and attention it was likely to elicit).

More generally, the above data show that, after the crisis, there was a significant rise in both the output of the four think tanks and in the coverage it received. Since most think tanks have as their mission to inform policy and the public debate, they certainly stand to gain from increasing media recognisability and attention. However, the access of different organisations to media outlets is uneven, and they faced challenges in expanding the reach of their ideas while controlling how they are interpreted. For instance, as NIESR's economic data became more politically central, it found difficulties in shaping how their message was reported. The crisis thus offered a privileged instance to expand NIESR's influence—as publics were avid for explanations and economic expertise—while entailing its own risks, in the form of challenges to their status as neutral arbiters.

However, public interventions cannot target the public as if it were a coherent whole. As John Dewey (1946 [1927]) suggests, in modern democracies, publics are disjointed—there is simply too much to pay attention to easily reach everyone. Dewey believed that the public is not a given, but in some sense has to be constructed; its very definition is an intellectual problem.[1] Perhaps that is why to whom think-tankers are speaking (and who is listening) is not always easy to answer. In some sense, think tanks both constitute and are constituted by their publics.

If we follow what is stated in most mission statements, the intended publics of think tanks are traditionally two: policymakers and the wider political debate—perhaps especially 'second-hand dealers in ideas': journalists, editors, teachers, civil society organisations, etc. Here the distinction between 'coordinative' and 'communicative' discourse (Schmidt 2008) is worth restating—between discourse aimed at the policymaking elite (policy reports, parliamentary hearings) and that meant for larger publics (op-eds, social media, etc.). Following Ladi (2011), I surmise that, although all four organisations engaged in both types of discourse, at least three factors determined the prominence of one type or the other, and hence of certain formats over others. First, a think tank's perceived closeness to policymakers. Second, the sources and modalities of their funding. Third, the emergence of junctures in which policy shifts are perceived to be more probable.

Regarding the first factor, already Denham and Garnett (1998) had noticed that most media-oriented think tanks tended to keep a certain

[1]"Indirect, extensive, enduring and serious consequences of conjoint and interactive behavior call a public into existence having a common interest in controlling these consequences. But the machine age has so enormously expanded, multiplied, intensified and complicated the scope of the indirect consequences [...] that the resultant public cannot identify and distinguish itself. [...] There are too many publics and too much of public concern for our existing resources to cope with. The problem of a democratically organized public is primarily and essentially an intellectual problem, in a degree to which the political affairs of prior ages offer no parallel". (Dewey 1946 [1927]: 126)

distance from government. Presumably, as a vigorous media strategy often places them on an often-antagonistic public debate, accusations of bias are likely, which in turn renders then unfit to be seen as impartial enough to directly inform policy. In Medvetz's terms, this is the trade-off between accumulating media and other types of capital (in this case political). Hence, think tanks situated at arm's length from party leaderships, such as the ASI, are more likely to generate considerable amounts of 'communicative discourse.' Similar reasons might also explain why, as the chasm between NIESR's recommendations and official policy grew, they produced ever more expert opinion through media channels—politics had drifted too far away from evidence to only be influenced through specialist advice. Conversely, as PX became ever more politically central, it produced ever more 'coordinative discourse'—for instance in the reports where they assessed in which government departments austerity should concentrate. PX's impressive level of production of policy blueprints in 2009–2010, as a new Conservative-led Coalition came to power, is another case in point.

Concerning the second factor influencing format, it is worth noting that the public interventions of think tanks are linked both to the volume and type of funding they pursue. As the contrast between NEF and the ASI shows, the regime under which think tanks are supported is critical. While project-based funding allows for greater production of in-house research, it often limits the time think-tankers have for other endeavours—i.e., setting out their own research priorities and intervening more actively through 'communicative discourse.' While NEF has around 50 employees, only James Meadway, and only towards the end of this book's timeframe, was predominantly tasked with writing media op-eds and blogs, compared to ASI's staff of 5–10, almost all of whom write frequently for a sympathetic press. That is, think tanks with significant core funding can easily produce quick-response commentary on the issues of the day, while those funded through research-contracts (i.e., most in Britain) are more likely to produce mid-range, medium-term, empirical work.

Another variable that determines the effect of funding on think tanks' public interventions is, to be sure, their provenance. Different organisational characteristics and political positions impact the type of funders an organisation is likely to attract. In Britain, most financial support for think tanks comes from charitable trusts, corporate donations, wealthy individuals, government departments, EU institutions, local government, trade organisations, and research councils. These constituencies vary both in the volumes of funding they provide and in their modality. Academic research grants, for instance, have different objectives, reputational effects, and bureaucratic obligations than most forms of private sector funding. Overall, from the four cases covered in this book, one could say that, with the exception of the ASI, support from charitable trusts is becoming more widespread, despite initial fears that it would shrink after the recession (Charity Commission 2010; Alcock et al. 2012). This tendency is well illustrated by NIESR and has important effects on the way think tanks are likely to present themselves, prioritising short-term work that maximises their perceived public benefit as defined by funders.

The third factor that affects the likely format of think tanks' public interventions is their 'timeliness.' Ladi (2011) claims 'communicative discourse' tends to be favoured in junctures where the policy debate is perceived as likely to shift; or 'windows of opportunity' (Kingdon 2003). This idea is linked to the significant role think tanks have been reported to have in defining the contours of policy problems (see Chapter 2). An index of their probable rise to centrality in such moments is the remarkable increase in their research output—which is linked to greater funding for work on macroeconomics and financial policy. Crises, in this sense, can open the space for new public interventions by boosting the demand for new expert narratives.

There are, however, at least three ways in which the distinction between 'communicative' or 'coordinative' discourse is problematic. The first I shall illustrate with economic indicators. As NIESR's experience shows, even technical interventions on narrow policy issues can become the subject of heated public deliberation. Hence, even if a public intervention is targeted at specialist audiences, some formats—particularly those involving numbers—can have a life of their own. Thus considered, NIESR's inability to control how their economic work was read might help explain their increased production of 'communicative discourse.' The second caveat is think tanks' role as hosts. For instance, PX saw rising media coverage as a regular venue for official announcements by Conservative politicians. Although not 'public interventions' in a strict sense, these events—at the crossroads between being 'coordinative' and 'communicative'—helped position PX in the press and in networks of political power, which in turn grants access for further institutional resources. Events are also illustrative of the third sense in which the distinction between 'coordinative' and 'communicative' discourse can be moot. If, as Dewey argued, publics themselves have to be constructed, they are not completely independent of the 'speech act' that addresses them. Think tanks, when most successful, can help in the construction of their own publics through their capacity to bring together and 'moderate' otherwise quite distinct sectors of society; say, much like PX put in contact sympathetic journalists, academics, and buddying Conservative politicians.

Be that as it may, although think tanks increased their media attention and output, in line with greater demand for policy narratives, the dissemination of their ideas faced substantial obstacles. Earlier in this book, I posited that experts themselves were widely seen to have failed in the run-up to the crisis, and catastrophically so. This crisis of confidence opened the space for challengers to produce alternative discourses on the economy, opening a space for heterodox economists on the left and radical free-marketeers on the right. Still, this also meant that the ideas of many experts could be safely dismissed. That is, a context of epistemic crisis limited the possibilities of most actors to be seen as having 'cognitive autonomy' and 'epistemic authority'—that is, to be perceived as knowledgeable and impartial enough to command authoritativeness across large audiences. In sum, even though the demand for expert knowledge increased, experts themselves became more vulnerable to being accused of partiality. NIESR's experience is a case in point as are, beyond the timeframe of this research, the debates around whether people are tired of experts in the run-up to the Brexit referendum.

However, even if in critical moments who should be trusted can become more open to debate (and reputations more volatile), the 'mainstream' can remain difficult to assail for those in the periphery. An interviewee at NEF mentioned the organisation had difficulties "crack[ing] the FT." This implies that NEF remained, regardless of the merit of their arguments, relatively marginal in important quarters of the economic debate, even at a time when alternative narratives were widely sought. However, by the same token, NEF staff were well positioned among the avant-garde of economic thinking—e.g., students and heterodox experts at 'Rethinking Economics.' On that note, I believe future research on think tanks could focus on their effects on fields other than politics, especially those in flux (for instance, the discipline of economics).

Crises, Tensions, Resolutions

The barriers think tanks confront when attempting to reach across audiences bring us to the second objection to the *hysteresis* hypothesis I stated earlier: namely, that it exaggerates the internal coherence of think tanks. Indeed, incoordination is particularly likely during crises for two reasons. Firstly, although the majority of think tanks can be characterised as being animated by more than the advancement of pure knowledge, they still need to be seen as evidence-based. It might seem quaint in 2019, but in the years around the New year era, a 'what works' discourse of policymaking was dominant and a reputation for being driven only by ideological commitments was unseemly, especially when seeking political centrality. Further, given the complex character of the economic crisis—and that, in its aftermath, experts themselves were found wanting—an acknowledgement of the insufficiency of our knowledge became common form. Being seen as changing one's mind was not as damning as would have been in other contexts. Interestingly, during interviews, many think-tankers were adamant in highlighting instances where they changed their views. Secondly, moments of intense political uncertainty make internal tensions more likely to manifest themselves. No political position is completely air-tight, and one's membership to broader intellectual coalitions and networks almost necessarily means dealing with allies with (at least slightly) different priorities and worldviews. The following paragraphs will elaborate on these for each of the four cases.

NEF, whose traditional constituencies include environmentalists, campaigning organisations, and heterodox economists, had limited access to the media outside the centre-left and became marginal in economic policy as the austerity discourse took hold. Their 'Green New Deal'—itself an attempt to solve the tension between Keynesianism and Environmentalism—became increasingly at odds with a public debate centred on the need for 'belt-tightening' (though it is noteworthy that this concept has become more popular recently, especially in the United States). They had gone through a window of opportunity that was fast closing. As NEF's 'coordinative discourse' seemed ineffective, it sought to produce more 'communicative discourse.' At the same time, this focus on appealing to the public and changing the basic tenets of the economic debate was rendered difficult by their dependence on short-term research contracts and their irregular access to a mainstream media mostly dominated by the right. Given the above, intellectual change was visible in NEF, rather than in the content of their public interventions, on how their arguments were to be presented to publics unlikely to readily accept their narrative. That is, they shifted from providing an optimistic message on the need for a new economy, which seemed out of kilter with the public mood, to putting together a coordinated front to combat the pervasiveness of the austerity discourse. The analysis of 'frames' and the establishment of NEON are efforts to organise a network to contest not only the evidence base of the Coalition's economic policies but also their underlying view of how the economy works.

Conversely, although the ASI applauded the Coalition government's focus on fiscal discipline, in the years leading up to the crisis they had a diminishing base and suffered some loss of relevance (Pautz 2012b). Facing better-resourced competition in a more crowded Conservative camp, ASI's policy proposals risked being sidelined by more slick competition and pigeonholed as too extreme. And while the crisis opened the possibility of expanding their donor base, the perceived failures of capitalism and the disrepute of libertarian arguments (seen as fanatical in some quarters) presented obstacles for the dissemination of their message, especially among the young. Perhaps for those reasons, after the arrival of Sam Bowman and a new cadre of free-marketeers, the ASI sought to expand their appeal. Hence, their promotion of 'Bleeding-heart Libertarianism' is partly an attempt to reach audiences beyond their traditional supporters. However, more traditional rights-based libertarian arguments remained important in ASI's orbit, which elicited some interesting internal debates.

NIESR's experience of the crisis was marked by its uncomfortable place between academic economics and economic policy. Substantial growth in the visibility of their econometric work between 2008 and 2010 opened a space for informing policymakers and the public. Yet, they also faced challenges in how others interpreted their research and how their numbers were cited. NIESR opposed the Coalition's fiscal measures, which they associated with a growing rift between official policy and the expert consensus, all while seeking to expand their funding sources beyond cash-strapped government departments. In a few words, they found themselves in a position where their numbers were widely cited but their advice rarely heeded. In that context, a change of mode of public engagement was likely, which meant a more active pursuit of communicative discourse, if also trying to maintain a reputation for non-partisanship and scientific rigour. However, although NIESR's efforts enhanced their visibility, this did not mean their economic ideas became dominant and even risked accusations of having become politicised.

Finally, the trajectory of PX was shaped by their place at the centre of networks of power, by their status as the privileged forum for centre-right policy thinking in the Cameron years. PX's closeness to Conservative modernisers meant that a radical shift in the political platform of a future Conservative government was bound to have significant effects on their organisation. Hence why, for PX, internal tensions were visible in their attempt to assess and revitalise the ideas behind the 'Compassionate Conservatism' agenda once they had been cast aside by the purported need for austerity. In this sense, as an interface of political elites and policy expertise, the modernising ideas that had been developed in the run-up to the 2010 election, and which initially gave this think tank its identity, became secondary to their new mission: devising a detailed plan to implement fiscal consolidation that could be seen as reasonable and fit for the political mainstream.

Overall, I surmise that however different were the trajectories of these think tanks, at least four constants are observable. Firstly, in all cases, transformations were sedimentary. Parts of each of these institutions continued to produce work which was at odds with some of their new public interventions. For instance, given the breadth of research produced by NIESR and the specialist character of their work, much of it continued to operate at a distance from the media in less politically salient topics. Or, given that ASI continued to accommodate a priori, rights-based libertarians after the arrival of consequentialists, internal tensions were quite noticeable. A case in point is the institution of the annual Ayn Rand Lecture in 2012, named after a famous right-wing thinker who spent her career arguing that concerning ourselves with the plight of the poor is objectionable.

Secondly, these intellectual changes were embodied by individuals, especially, for our cases, Carys Afoko and Daniel Vockins, Sam Bowman, Jonathan Portes, and Andrew Lilico—for whatever reason, almost all white men, and all graduates from elite universities. After all, public interventions are always produced and presented by actual people. Think tanks do not speak for themselves but need someone to do it on their 'behalf.' Ideas, after all, cannot be easily detached from the people who convey them. Following this point, the continued association of particular individuals with these organisations can dramatically shape the format and content of their public interventions, and hence their position in the public imagination. Appreciating to a greater extent this almost self-evident fact could significantly enhance our understanding of how intellectual change in organisations comes about.

Thirdly, processes of repositioning were often shaped by the internal tensions underpinning each organisation. NEF's most salient was expressed in their indebtedness to the ideas of both Keynes and Schumacher (for pro-growth public spending vis-à-vis 'small is beautiful' environmentalism); ASI's was to be found in the theoretical opposition between consequentialist and a priori arguments for free markets; NIESR's in their orientation—either aloof or engaged—towards the policy debate; and PX's in their advocacy for both 'Compassionate Conservatism' and stringent public spending cuts. Nevertheless, each think tank, whenever possible, would attempt to supersede these internal strains. For instance, NEF's 'Green New Deal' sought to render compatible a need for economic stimulus with environmental concerns and PX's 'Digital Government' unit generated proposals that sought to both improve policy outcomes, engage the public, and cut government spending.

Fourthly, repositioning was more likely to occur when think-tankers became particularly concerned with the obstacles faced in the dissemination of their message. NEF and ASI's efforts to reach beyond their usual publics are the clearest example of this, but one could also mention NIESR's production of YouTube videos for non-specialists. Delving deeper in the tension between a think tank's message and its reception, in the next section I posit that zeroing in on their 'moderation' of many, often conflicting, publics and fields can help in deciphering the role of these organisations in modern politics.

Think Tanks as Moderators

Following this overview, I propose that the function of think tanks in the policy debate is best encapsulated by the word 'moderation,' which I use here mostly as a verb, if also hinting at the noun—they are, to be sure, not unrelated. After all, as most public debates on policy matters are presented as conflicts between ideological poles, what is perceived as reasonable often coincides with the 'moderate'—with a 'somewhere in between' two extremes. Think tanks that are perceived as radical might alter the contours of the debate and the bounds of the acceptable (as the ASI did in the 1980s), but political centrality often requires a reputation for pragmatism and temperance (Ashforth and Gibbs 1990). In this sense, most think tanks profit from being seen as authoritative as widely as possible, which requires reaching out to audiences who might not share their views. The aim of most think tanks, from this standpoint, would be to position themselves as 'reasonable' across publics, to define what that means, and—if acting as 'hosts' through their events, and publications—to 'moderate' what are legitimate contributions to the public debate.

The concept of moderation, perhaps due to its ambiguity and opacity, has been the object of little but growing attention in sociology (Holmwood 2013; Holmwood et al. 2013). However, if tacitly and vaguely, 'moderation' (both as verb and noun) is not an uncommon way of understanding the role of think tanks and experts. Many have claimed that think tanks seek to 'mediate' (Osborne 2004) and be seen as 'bridges' connecting relevant socio-scientific research with politics and policy (Stone 2007). A similar idea is implicit in Medvetz's (2012a) model of think tanks as boundary organisations that convert capitals from one social field to another. In another context, Silva (2009: 18) spoke of experts as a 'moderating force' between antagonistic political actors who would otherwise engage in frontal conflict. And in the United States, right-wing think tanks have commonly portrayed themselves as a response to the dominance of the left in academia, seeking to balance the public debate by providing extra-mural sources of expert knowledge (Stahl 2016).

Critics have rightly stressed that think tanks are not impartial mediators of expert knowledge, and that what the knowledge they select as relevant for policy is neither neutral nor casual (Stone, op. cit.; Kay et al. 2013). Others have objected that the equivalence between right-leaning think tanks and left-of-centre university departments is misleading (Stahl, op. cit.). Agreeing with these caveats, I shall attempt, rather than to discuss whether think tanks are proper conduits of research into policy in an 'enlightenment model' of policy influence, or whether they actually 'balance' the debate, to examine how they *perform* moderation. In other words, think tanks, where most successful, can help define the contours of what is seen as worthy of attention in the policy debate, shaping a common ground that establishes the limits of the conversation across audiences.

Nevertheless, much like political extremes are not fixed, there is no objective, stable benchmark to settle what should be considered moderate. Its definition is, by necessity, subjective, relational, contextual, and political. Moderation, as a perceived political and intellectual position that rejects but mediates between extremes, is perhaps particularly weighty in Westminster, with its political tradition greatly indebted to Fabians on the left and Burkeans on the right. Further, in the two decades coming to the crisis, British mainstream parties—first Labour (Pautz 2012a) and later the Conservatives (Wade 2013)—famously sought to capture a chimerical political centre, inspired by the policy proposals of 'modernising' think tanks such as IPPR and PX. This was underpinned by a general diagnostic that a Manichean approach to policy had become politically unwise and unattractive after the end of the Cold War. Ideological purity was, at least discursively, seen as a weakness as a 'what works' discourse sprung and became dominant, at least for a while. In this milieu, a new generation of think-tankers arose, which sought to widen the appeal of their political sector while devising policies that could be deemed workable, reasonable, and politically agreeable. They configured a new 'wave' of think tanks that distinguished itself from the more ideologically-driven institutions of the 1970s.

However, the above is not to equate 'moderation' with 'centrism.' Whereas the latter is mostly associated with tepid inconsistency (Holmwood et al. 2013: 8), moderation requires a minimum perceived coherence and the bridging of demands from across publics. For that reason, definitions of the moderate and the reasonable abound in the think tank world. Most think-tankers consider themselves moderate in some broad sense, or at least claim they do not subscribe to an unwavering 'dogma.' NIESR staff members definitely would not, as they seek to be perceived as driven by economic evidence; nor NEF's, who aim to provide innovative thinking, unburdened by 'old' economics, to tackle impending and unprecedented crises in the economy, society, and the environment. To be sure, PX members do not define themselves through the adherence to a dogma either, as their professed goal was to produce sound and politically feasible policies for a modernised centre-right. Even the ASI—perhaps among our case-studies the most clearly and combatively positioned at one end of the spectrum—expanded its strategy beyond trying to move the 'Overton window' by pulling it from one extreme. At least since the arrival of Bleeding-heart Libertarians, the ASI has sought to be seen as evidence-informed and to appeal to audiences larger than card-carrying free-marketeers. Regardless, if one were to ask members of each of these organisations about the others, I wager most would consider the others quite far from 'moderate'. A few interviewees sniggered after, when asked, I mentioned which other think tanks were part of this study.

The above begs the questions: who decides what is moderate? And under what circumstances can it change? Furthermore, writing in the aftermath of the Brexit referendum and the electoral victories of the likes of Trump and Bolsonaro, one could reasonably claim that modern politics has ceased to be driven by moderates, and we have entered a new age of extremes. There are at least two ways of addressing these objections. The first would be to agree and submit as proof the growing irrelevance of think tanks and policy experts in the aftermath of the rise of populists in the USA, Europe, and elsewhere (see *Washington Post* 2017). There is some truth to this assertion, though one ought to avoid thinking that think tanks *are* moderate in some intrinsic, immutable sense. A second way would be to claim that political and epistemic crises are bound to affect who is allowed to *moderate* in the first place.

Arguably, in 'normal times'—those not widely perceived as requiring urgent action—there is an elective affinity between stasis and political moderation, and hence between the powers that be and what is considered reasonable and even-handed. In other words, in such times the spaces for moderation and encounter across publics are relatively stable. However, when a moment is construed as critical, political action becomes a necessity and the centre cannot hold. Discourses on crises are thus of the utmost political importance: crises differentiate epochs and elevate purposive political action with uncertain consequences to the centre-stage. For that reason, the notion of crisis has been used in historical research in at least two interrelated senses (see Koselleck 2002: 240). First, as an 'iterative periodising concept,' as watersheds where history could have gone one way or another; Thucydides is among the first to have used the word in relation to some crucial battles of the Peloponnesian wars. Second, as the judgement of the world, the unravelling of what was previously thought of as settled, which shows its underlying truth—often meaning the end of days. Hence, if a juncture is thought to be 'critical,' a space opens to redefine what course of action is reasonable and what is worthy of attention.

The events of 2008 are a good illustration of this process. The global financial crisis undermined the status of academic economists in the public imagination, rendering more uncertain our view of their subject matter. From that point on, it became less unreasonable to be suspicious of the pronouncements of these experts. What was hitherto believed to be solid, the primacy of mainstream economics in policymaking and in circles of power, melted into air. However, and enigmatically, this crisis of the authority of economists occurred in parallel to greater demand for decisive political action over the economic sphere and the emergence of austerity, a policy discourse that presented itself as strictly driven by economic considerations. One could say that after the crisis the 'economy' became a spectre that could act on its own, without the need for a disavowed economic consensus—like a phantom limb.

Authors working on think tanks from across theoretical traditions—be they inspired by North American political science (Abelson 2012), Gramsci (e.g., Pautz 2016), or discursive institutionalism (Ladi 2011)—have argued that think tanks are most likely to be influential in moments when a political shift seems likely and the contours of a situation need to be defined. The aftermath of 2008 was, undoubtedly, such a moment. This chimes with the sense of opportunity and danger many think-tankers expressed in the months after the crisis. And though one should not exaggerate the importance of these organisations as drivers of policy—they operate in a crowded space, face many external pressures, and were undergoing critical processes of their own—the situation gave the opportunity to well-positioned think-tankers to, through their role as moderators, help set up a 'climate' in which ideas to read and address the crisis could take shape.

In the UK, as the argument that fiscal discipline was a necessity became widespread (Berry 2016), much of the political debate centred on what was a reasonable level of government debt. While in early 2009 Conservatives matched Labour's spending plans, later that year they began to claim that fiscal profligacy had brought about the recession and that it was not only sensible, but also a necessity, to implement severe cuts. A 'moderation' of public spending was often rhetorically opposed to the 'excess' of the Labour years (de Goede 2009; Sinclair 2010). It follows that the radicality of the policies that were eventually implemented did not preclude them from being perceived as moderate. In that sense, the aim of those pushing for austerity was to be seen as practical rather than ideological by a majority of the public, even if going against the views of most policy experts. These policies, one ought not to forget, would have been considered inadmissibly drastic by most politicians before 2009.

Looking at the experience of NIESR prompts one to ask what the role of technocratic specialists is in the moderation of fields. If we follow their advice, the official policy response to the recession was unreasonable, unencumbered by any consideration of the economic consensus. But that consensus had ceased to be considered reliable and trustworthy. To put it in perhaps coarse Bourdieusian terms, the 'capital conversion rate' of academic economics had sunk, and a view that was marginal in academia could become central in politics and the media. Hence why I posit that the study of the emergent interstitial field of think tanks that Medvetz (2012b) proposes should centre on the conditions of possibility for such organisations to 'moderate' across fields—and thus, if convenient, take minority positions from within one to justify an action in another. In that way, following Medvetz, moderation should be considered in relation to the wider dynamics of fields, highlighting how what is considered worthy beyond its field of provenance is negotiated in an unstable environment. By way of illustration, this can mean that what is broadly seen as economically moderate and reasonable is not necessarily determined by economists.

Among our four think tanks, PX is the most likely candidate to fit the description of moderation outlined here. Nevertheless, its position at the centre of networks of power carried its own risks, as it meant keeping apace with a volatile political environment. While 'Compassionate Conservatism' had defined PX in its early years, its vocation for policy impact required for that project to be abandoned when it was unpropitious. Thus considered, even the most influential of think tanks can be shaped by external forces. PX also had to move to still be able to 'moderate,' which is also noticeable in the hardening of their views, as years went by, on economics, social policy, the UK's EU membership, and immigration. As shown throughout this book, political centrality often necessitates a malleable intellectual disposition, but which causes which is another matter. After all, what is moderate is always defined in relation to the 'extremes' that are on the table. In this sense I believe we should understand Holmwood's warnings. When discussing the relationship between experts and the public, he claimed:

> [W]here expertise is in the service of political or administrative elites it is likely to be vulnerable to populist mobilizations by the very interests that expert opinion is being called upon to moderate. (Holmwood 2013: 187)

Writing in 2019, the mistrust of expertise has become ever more pervasive—even if some form of expert discourse is always needed for the substantiation and legitimation of politics. In this context, the specialised experts of yesterday ceased to be considered moderate, those seeking to speak *truth* to power from a position of traditional expertise faced ever-greater challenges to be taken seriously. In their place, a new generation of 'bounded innovators' sprung whose mission is to furnish 'politically-fit' policy expertise. Think tanks help construct their publics, but are also constructed by them.

References

Abelson, D. (2012). Theoretical models and approaches to understanding the role of lobbies and think tanks in US foreign policy. In S. Brooks, D. Stasiak, & T. Zyro (Eds.), *Policy expertise in contemporary democracies*. Farnham: Ashgate.

Alcock, P., Parry, J., & Taylor, R. (2012). *From crisis to mixed picture to phoney war: Tracing third sector discourse in the 2008/9 recession* (Third Sector Research Centre Research Report (78)). Accessed 27 May 2015. http://epapers.bham.ac.uk/1780/1/RR78_From_crisis_to_mixed_picture_to_phoney_war_%2D_Taylor%2C_Parry_and_Alcock%2C_April_2012.pdf.

Ashforth, B., & Gibbs, B. (1990). The double-edge of organizational legitimation. *Organization Science, 1*(2), 177–194.

Berry, M. (2016). No alternative to austerity: How BBC broadcast news reported the deficit debate. *Media, Culture and Society, 38*(6), 844–863.

Charity Commission. (2010). *Charities and the economic downturn*. Accessed 18 November 2015. https://www.gov.uk/government/publications/charities-and-the-economic-downturn-parliamentary-briefing.

de Goede, M. (2009). Finance and the excess: The politics of visibility in the credit crisis. *Zeitschrift für Internationale Beziehungen, 16*(2), 295–306.

Denham, A., & Garnett, M. (1998). *British think tanks and the climate of opinion*. London: UCL Press.

Dewey, J. (1946 [1927]). *The Public and its problems: An essay in political inquiry*. Chicago: Gateway Books.
Holmwood, J. (2013). Rethinking moderation in a pragmatist frame. *The Sociological Review, 61*(2), 180–195.
Holmwood, J., Smith, T., & Thomas, A. (2013). Sociologies of moderation. *The Sociological Review, 61*(2), 6–17.
Kay, L., Smith, K., & Torres, J. (2013). Think tanks as research mediators? Case studies from public health. *Evidence and Policy, 59*(3), 371–390.
Kingdon, J. (2003). *Agendas, alternatives and public policies*. New York: Longman.
Koselleck, R. (2002). *The practice of conceptual history*. Stanford: Stanford University Press.
Ladi, S. (2011). Think tanks, discursive institutionalism and policy change. In G. Papanagnou (Ed.), *Social science and policy challenges: Democracy, values and capacities*. Paris: UNESCO.
Medvetz, T. (2012a). *Think tanks in America*. Chicago: University of Chicago Press.
Medvetz, T. (2012b). Murky power: 'Think tanks' as boundary organizations. In D. Golsorkhi, D. Courpasson, & J. Sallaz (Eds.), *Rethinking power in organizations, institutions, and markets: Research in the sociology of organizations* (pp. 113–133). Bingley: Emerald Group Publishing.
Osborne, T. (2004). On mediators: Intellectuals and the ideas trade in the knowledge society. *Economy & Society, 33*(4), 430–447.
Pautz, H. (2012a). *Think tanks, social democracy and social policy*. London: Palgrave Macmillan.
Pautz, H. (2012b). The think tanks behind 'cameronism'. *British Journal of Politics and International Relations, 15*(3), 362–377.
Pautz, H. (2016). Managing the crisis? Think tanks and the British response to global financial crisis and great recession. *Critical Policy Studies, 11*(2), 191–210 [Online early access].
Schmidt, V. (2008). Discursive institutionalism: The explanatory power of ideas and discourse. *Political Science, 11*(1), 303–322.
Silva, P. (2009). *In the name of reason: Technocrats and politics in Chile*. University Park: Penn State University Press.
Sinclair, T. (2010). Round up the usual suspects: Blame and the subprime crisis. *New Political Economy, 15*(1), 91–107.
Stahl, J. (2016). *Right moves: The conservative think tank in American political culture since 1945*. Chapel Hill: University of North Carolina Press.

Stone, D. (2007). Recycling bins, garbage cans or think tanks? Three myths regarding policy analysis institutes. *Public Administration, 85*(2), 259–278.
Wade, R. (2013). *Conservative Party economic policy: From Heath in opposition to Cameron in coalition*. Basingstoke: Palgrave Macmillan.
Washington Post. (2017). *Trump could cause 'the death of think tanks as we know them'*. Accessed 12 May 2017. https://www.washingtonpost.com/opinions/global-opinions/trump-could-cause-the-death-of-think-tanks-as-we-know-them/2017/01/15/8ec3734e-d9c5–11e6-9a36-1d296534b31e_story.html?utm_term=.48b548c7a5f9.

Afterword: For a Comparative Sociology of Intellectual Change

The underlying theoretical motivations behind this book were, above all, two. First, I was interested in how agents with an intellectual mission react to crises with an epistemic element; when stakes are high and they have a material investment in the ideas they advocate. Given the importance of the financial industry in the UK, British think tanks after the 2008 crisis were a particularly appropriate proxy to detect how such a type of intellectual change occurs. After having covered their experience, I hope to have contributed to a research programme to study intellectual coordination and instability. I hope to have provided, in however vague and humble a fashion, readers with a theory and methods appropriate to examine intellectual collectives, their public interventions, their internal coordination, and how they interact with their environment.

The second purpose, to which I dedicate this book's last paragraphs, concerns the role of sociologists as 'second-order' intellectuals, or our pretension to be so. Many in our discipline too often fail to notice that, whatever the promise of sociology for understanding knowledge in society, its authority has been eroded across much of the public, maybe in

M. González Hernando, *British Think Tanks After the 2008 Global Financial Crisis*, Palgrave Studies in Science, Knowledge and Policy, https://doi.org/10.1007/978-3-030-20370-2

some parts of it beyond repair. Whether by our own doing or not, sociologists are far too frequently considered not impartial enough to have expert authority. Hence, outside sociologists' usual audiences—and in an increasingly anti-intellectual climate across much of the world—many do not wish to hear of sociology's concepts and findings, not (only) because they are assumed to be uninteresting or unreliable, but because they are presumed to be politically suspect.

In the case of sociology and anthropology, the historical advantage of abandoning the ideals of objectivity and of becoming positive sciences was a deeper awareness of researchers' own position *vis-à-vis* their 'object of study.' That was doubtlessly commendable and necessary. But along with it came an abdication: Why should social scientists be listened to any more than others passing judgement on social issues? Hence why so many of our public interventions are stillborn. Hence why others, among them deft and partisan think-tankers, have taken the place in the public debate many sociologists imagined for themselves (Medvetz 2012; Misztal 2012).

It is not true, however, that these suspicions are solely a concern for 'soft' disciplines. Nowadays not even the 'hard' sciences—or those aspiring to such a status, as NIESR's experience suggests—command the authority they once did. Moreover, even putting aside the 2008 financial crash, the reputation of those who seek to be seen as experts to credibly describe and foresee the world has come increasingly under question. This is the case, for instance, in current debates over anthropogenic climate change and the safety of MMR vaccines. In the void left by traditional epistemic authorities, which could no longer simply dismiss outsiders by calling them quacks, new intervenors have appeared who need not be constrained by expert consensus. In the think tank world, this has created a space for the proliferation of 'apposite' policy ideas and expertise, for which cognitive autonomy and convincing specialist peers are only of secondary importance.

In the long run, this mistrust of experts has come to affect think-tankers themselves, even (perhaps particularly) those most politically influential. Policy institutes whose brands hinge on centrality are faced with the prospect of having their role restricted to translating political pressures into a workable policy form. As one interviewee put it, in the future the function of think tanks could become simply to "jump on the bandwagon" (ASI interview), following politics rather than informing it. To put it bluntly, the situation is propitious for the emergence of 'moderate' think-tankers tasked with giving a veneer of expertise to ultimately unreasonable policies. After 2016, one does not need to think much to come up with examples of this risk.

In view of the above, one direction this project could have taken would have consisted in castigating most think-tankers as *ersatz* experts, hacks, mouthpieces. Under a normative definition of intellectuals *à la* Saïd (1994), such a conclusion would have been easy to reach, and indeed I must say at times I was tempted. I have refused it on methodological and pragmatic grounds: think tanks produce relevant public interventions that traverse publics, and hence their part in the public debate is necessary to observe regardless of their perceived biases. Furthermore, as I hope to have shown, think tanks show an impressive capacity to change and learn and, whatever the reasons driving that change, we can learn much by observing that.

But there is also another, more profound reason to avoid simply dismissing think tanks as epiphenomena. Many sociologists writing on think tanks have complained that their effect is to 'crowd out' more autonomous intellectuals, generally implying academics. However, when mistrust of expertise has become commonplace, this type of jeremiad does not achieve much. It becomes merely a complaint from actors perceived to be just as biased, and always one step behind those they critique.

Instead of making that protestation, I posit that a sociologyof intellectuals could make a crucial contribution to go beyond the stale divides between objectivity and subjectivity and between experts, propagandists, and laypeople. But this sociology would demand an uncanny estrangement. It would require an appetite for vicariousness and an interpretive bent, already noticeable in the Weberian tradition, if not to access the worldviews of others, at least to attempt to understand the logic behind what they say and how what they say makes sense at a particular point in time and space. Such an effort, which by necessity would be comparative, would require us to learn to speak like those with whom, at a political level, we vehemently disagree with.

However, in order to command the respect needed to be taken seriously in that exercise, we would have to make a much greater effort to learn others'—several others'—languages. We would have to think in their own terms, minding the fact that ideas have roots and weight. More fundamentally still, we would have to think we can experience the world like others—even partisan think-tankers—to, ultimately, understand the context in which ideas that seem pernicious to us are deemed convincing. Then would come the hard work of telling the story of the stories we tell ourselves and each other, to show how they evolve and how they are connected with the position from which we speak. In short, we would have to try to do what Akira Kurosawa did in *Rashomon*.

I sought to contribute something along those lines, in the understanding that a sociology of intellectuals is, ultimately, a sociology of our polity.

References

Medvetz, T. (2012). *Think tanks in America*. Chicago: University of Chicago Press.

Misztal, B. (2012). Public intellectuals and think tanks: A free market in ideas? *International Journal of Politics, Culture and Society, 25*(4), 127–141.

References

Abelson, D. (2002). Do think tanks matter? Opportunities, constraints and incentives for think tanks in Canada and the United States. *Global Society, 14*(2), 213–236.

Abelson, D. (2012). Theoretical models and approaches to understanding the role of lobbies and think tanks in US foreign policy. In S. Brooks, D. Stasiak, & T. Zyro (Eds.), *Policy expertise in contemporary democracies*. Farnham: Ashgate.

Aberbach, J., & Rockman, B. (2002). Conducting and coding elite interviews. *PS: Political Science & Politics, 35*(4), 673–676.

Abolafia, M. (2010). Narrative construction as sensemaking: How a central bank thinks. *Organization Studies, 31*(3), 349–367.

Ahearne, J. (2010). *Intellectuals, culture and public policy in France: Approaches from the left*. Liverpool: Liverpool University Press.

Alcock, P., Parry, J., & Taylor, R. (2012). *From crisis to mixed picture to phoney war: Tracing third sector discourse in the 2008/9 recession* (Third Sector Research Centre Research Report [78]). Accessed 27 May 2015. http://epapers.bham.ac.uk/1780/1/RR78_From_crisis_to_mixed_picture_to_phoney_war_%2D_Taylor%2C_Parry_and_Alcock%2C_April_2012.pdf.

M. González Hernando, *British Think Tanks After the 2008 Global Financial Crisis*, Palgrave Studies in Science, Knowledge and Policy, https://doi.org/10.1007/978-3-030-20370-2

Alesina, A., & Ardagna, S. (2009). *Large changes in fiscal policy: Taxes versus spending* (NBER Working Papers, 15438). Accessed 10 April 2015. http://www.nber.org/papers/w15438.pdf.

Alonso, W., & Starr, P. (Eds.). (1987). *The politics of numbers*. New York: Sage.

Álvarez-Rivadulla, M., Markoff, J., & Montecinos, V. (2010). The trans-American market advocacy think tank movement. In A. Garcé & G. Uña (Eds.), *Think tanks and public policies in Latin America* (pp. 172–199). CIPPEC: Buenos Aires.

Anderson, R. (2016). The Rashomon effect and communication. *Canadian Journal of Communication, 41*(2), 250–265.

Ashforth, B., & Gibbs, B. (1990). The double-edge of organizational legitimation. *Organization Science, 1*(2), 177–194.

Atkinson, P., & Coffey, A. (2004). Analysing documentary realities. In D. Silverman (Ed.), *Qualitative research: Theory, method and practice* (pp. 56–75). London: Sage.

Aupers, S. (2012). Trust no one: Modernization, paranoia and conspiracy culture. *European Journal of Communication, 27*(22), 22–34.

Austin, J. (1961). *How to do things with words*. Oxford: Clarendon.

Babrow, A., & Russell, L. (2011). Risk in the making: Narrative, problematic integration, and the social construction of risk. *Communication Theory, 21*(3), 239–260.

Baert, P. (2012). Positioning theory and intellectual interventions. *Journal for the Theory of Social Behaviour, 42*(3), 304–324.

Baert, P. (2015). *The existentialist moment: The rise of Sartre as a public intellectual*. Cambridge: Polity.

Baert, P. (2016). The philosopher as public intellectual. In M. Desch (Ed.), *Public intellectuals in the global arena: Professors or pundits* (pp. 163–181). Notre Dame: University of Notre Dame Press.

Baert, P., & Booth, J. (2012). Tensions within the public intellectual: Political interventions from Dreyfus to the new social media. *International Journal of Politics, Culture and Society, 25*(4), 111–126.

Baert, P., & Morgan, M. (2015). *Conflict in the academy: A study in the sociology of intellectuals*. London: Palgrave Pivot.

Baert, P., & Shipman, A. (2012). Transforming the intellectual. In P. Baert & F. Domínguez Rubio (Eds.), *The politics of knowledge* (pp. 179–204). New York: Routledge.

Ball, S., & Exley, S. (2010). Making policy with 'good ideas': Policy networks and the 'intellectuals' of New Labour. *Journal of Education Policy, 25*(2), 151–169.

Banet-Weiser, S. (2012). Branding the crisis. In J. Caraça, G. Cardoso, & M. Castells (Eds.), *Aftermath: The cultures of economic crisis* (pp. 107–131). Oxford: Oxford University Press.

Barrell, R., Davis, P., Karim, D., & Liadze, I. (2010). Bank regulation, property prices and early warning systems for banking crises in OECD countries. *Journal of Banking and Finance, 34,* 2255–2264.

Barrell, R., Hurst, I., & Kirby, S. (2009). The macroeconomic implications of pension reform. In D. France (Ed.), *Pension reform, fiscal policy and economic performance.* Rome: Bank of Italy.

Bauman, Z. (1989). *Legislators and interpreters: On modernity, postmodernity and intellectuals.* Cambridge: Polity.

BBC. (2007). *Northern Rock: Expert views.* Accessed 20 March 2015. http://news.bbc.co.uk/1/hi/business/6999246.stm.

BBC. (2008). *Policy Exchange dispute—Update.* Accessed 15 February 2016. http://www.bbc.co.uk/blogs/newsnight/2008/05/policy_exchange_dispute_update.html.

BBC. (2009a). *Brown calls for 'green new deal.'* Accessed 10 October 2014. http://news.bbc.co.uk/1/hi/uk_politics/7927381.stm.

BBC. (2009b). *UK recession 'worst since 1930s.'* Accessed 28 October 2015. http://news.bbc.co.uk/1/hi/business/8034879.stm.

BBC. (2009c). *UK recovery 'to take five years.'* Accessed 28 October 2015. http://news.bbc.co.uk/1/hi/business/8162217.stm.

BBC. (2013). *Eamonn on BBC Breakfast TV on fizzy drinks tax.* Accessed 21 March 2015. https://www.youtube.com/watch?v=YWIyjUNlZ08.

BBC. (2017). *Crash was economists' 'Michael Fish' moment, says Andy Haldane.* Accessed 15 September 2018. https://www.bbc.com/news/uk-politics-38525924.

Beck, U., & Wehling, P. (2012). The politics of non-knowing: An emergent area of social and political conflict in reflexive modernity. In P. Baert & F. Domínguez Rubio (Eds.), *The politics of knowledge* (pp. 33–57). New York: Routledge.

Benson, R. (1999). Field theory in comparative context: A new paradigm for media studies. *Theory & Society, 28*(3), 463–498.

Bentham, J. (2006). The IPPR and Demos: Think tanks of the new social democracy. *Political Quarterly, 77*(2), 166–174.

Berry, M. (2016). No alternative to austerity: How BBC broadcast news reported the deficit debate. *Media, Culture & Society, 38*(6), 844–863.

Bloor, D. (1991 [1976]). *Knowledge and social imagery*. Chicago: University of Chicago Press.

Bockman, J., & Eyal, G. (2002). Eastern Europe as a laboratory for economic knowledge: The transnational roots of neoliberalism. *American Journal of Sociology, 108*(2), 310–352.

Boin, A., t' Hart, P., & McConnell, A. (2009). Crisis exploitation: Political and policy impacts of framing contests. *Journal of European Public Policy, 16*(1), 81–106.

Boje, D. (2001). *Narrative methods for organizational and communication research*. London: Sage.

Boltanski, L., & Thévenot, L. (2006). *On justification: Economies of worth*. Princeton: Princeton University Press.

Boswell, C. (2008). The political functions of expert knowledge: Knowledge and legitimation in European Union immigration policy. *Journal of European Public Policy, 15*(4), 471–488.

Boswell, C. (2009). *The political uses of expert knowledge: Immigration policy and social research*. Cambridge: Cambridge University Press.

Bourdieu, P. (1977). *Outline of a theory of practice*. Cambridge: Cambridge University Press.

Bourdieu, P. (1984). *Distinction: A social critique of the judgement of taste*. Cambridge, MA: Harvard University Press.

Bourdieu, P. (1986). The forms of capital. In J. Richardson (Ed.), *Handbook of theory and research for the sociology of education* (pp. 241–258). New York: Greenwood.

Bourdieu, P. (1988). Vive la crise!: For heterodoxy in social science. *Theory & Society, 17*(5), 773–787.

Bourdieu, P. (1990). *Homo academicus*. Stanford: Stanford University Press.

Bourdieu, P. (1991). *Language and symbolic power*. Cambridge, MA: Harvard University Press.

Bourdieu, (1996 [1992]). *The rules of art: Genesis and structure of the literary field*. Stanford: Stanford University Press.

Bourdieu, P. (2000). *Les structures sociales de l'économie*. Paris: Éditions du Seuil.

Bourdieu, P. (2001). *Science de la science et réflexivité*. Paris: Raisons d'agir.

Bourdieu, P., & Wacquant, L. (1992). *An invitation to reflexive sociology*. Chicago: University of Chicago Press.

Bowen, G. (2009). Document analysis as a qualitative research method. *Qualitative Research Journal, 9*(2), 27–40.

Boyle, D. (2014). *Broke: Who killed the middle classes?* London: Fourth Estate.

Brändström, A., & Kuipers, S. (2003). From 'normal incidents' to political crises: Understanding the selective politicization of policy failures. *Government and Opposition, 38*(3), 279–305.

Brenton, M., Maltby, K., & Shorthouse, R. (Eds.). (2014). *The moderniser's manifesto*. London: Bright Blue.

Bronk, R. (2009). *The romantic economist: Imagination in economics*. Cambridge: Cambridge University Press.

Brooks, S. (2012). Speaking truth to power: The paradox of the intellectual in the visual information age. In S. Brooks, D. Stasiak, & T. Zyro (Eds.), *Policy expertise in contemporary democracies* (pp. 69–85). Farnham: Ashgate.

Bryan, D., Martin, R., Montgomerie, J., & Williams, K. (2012). An important failure: Knowledge limits and the financial crisis. *Economy & Society, 41*(3), 299–315.

Budd, A. (1999). Learning from the wise people. *The Manchester School Supplement, 67*(Suppl.), 36–48.

Burt, R. (2005). *Brokerage and closure: An introduction to social capital*. Oxford: Oxford University Press.

Business Spectator. (2010). *Oliver Marc Hartwich: It's time Henry had an umpire*. Accessed 3 April 2016. https://web.archive.org/web/20120415235526/http://www.businessspectator.com.au/bs.nsf/Article/MRRT-RSPR-Treasury-statistics-macroeconomics-pd20100720-7J7PX?OpenDocument.

Butler, E. (1985). *Hayek: His contribution to the political and economic thought of our time*. New York: Universe Publishing.

Butler, E. (2007). *Adam Smith: A primer*. London: IEA.

Butler, E. (2008). *The best book on the market: How to stop worrying and love the free economy*. Oxford: Capstone.

Butler, E. (2009a). *The rotten state of Britain: Who is causing the crisis and how to solve it*. London: Gibson Square.

Butler, E. (2009b). The financial crisis: Blame governments, not bankers. In P. Booth (Ed.), *Verdict on the crash: Causes and policy implications* (pp. 51–58). London: IEA.

Butler, E. (2010a). *The alternative manifesto: A 12-step programme to remake Britain*. London: Gibson Square.

Butler, E. (2010b). *Ludwig Von Mises: A primer*. London: IEA.

Butler, E. (2010c). *Austrian economics: A primer*. London: IEA.

Butler, E. (2011a). *Milton Friedman: A concise guide to the ideas and influence of the free-market economist*. Petersfield: Harriman House.
Butler, E. (2011b). *The condensed wealth of nations*. London: Adam Smith Institute.
Butler, E. (2012a). *Friedrich Hayek: The ideas and influence of the libertarian economist*. Petersfield: Harriman House.
Butler, E. (2012b). *Public choice: A primer*. London: IEA.
Butler, E. (2013). *Foundations of a free society*. London: IEA.
Calhoun, C. (1995). *Critical social theory: Culture, history and the challenge of difference*. Oxford: Blackwell.
Callon, M. (2007). What does it means to say that economics is performative? In D. MacKenzie, F. Muniesa, & L. Siu (Eds.), *Do economists make markets? On the performativity of economics* (pp. 311–356). Princeton, NJ: Princeton University Press.
Campbell, J. (2002). Ideas, politics and public policy. *Annual Review of Sociology, 28*(1), 21–38.
Campbell, J., & Pedersen, O. (2011). Knowledge regimes and comparative political economy. In D. Béland & R. Cox (Eds.), *Ideas and politics in social science research* (pp. 167–190). New York: Oxford University Press.
Carstensen, M. (2011). Paradigm man vs. the bricoleur: Bricolage as an alternative vision of agency in ideational change. *European Political Science Review, 3*(1), 147–167.
Charity Commission. (2010). *Charities and the economic downturn*. Accessed 18 November 2015. https://www.gov.uk/government/publications/charities-and-the-economic-downturn-parliamentary-briefing.
Charity Commission. (2013, September 1). *What makes a charity (CC4)*. Accessed 10 January 2016. https://www.gov.uk/government/publications/what-makes-a-charity-cc4.
Clark, S. (2012). *They can't help themselves: Anyone who contradicts the tobacco control industry must be a stooge of Big Tobacco*. London: FOREST. Accessed 15 March 2015. http://taking-liberties.squarespace.com/blog/2012/2/20/asi-acting-as-mouthpiece-for-the-tobacco-industry-says-ash.html.
Coats, A. W. (Ed.). (2000). *The development of economics in Western Europe since 1945*. London: Routledge.
Cockett, R. (1995). *Thinking the unthinkable: Think tanks and the economic counter-revolution 1931–1983*. London: HarperCollins.
Collini, S. (2006). *Absent minds: Intellectuals in Britain*. Cambridge: Cambridge University Press.

Collins, R. (2002). *The sociology of philosophies*. Cambridge, MA: Belknap Press.

ConservativeHome. (2009). *Policy Exchange appoints Andrew Lilico as its chief economist*. Accessed 12 March 2016 http://conservativehome.blogs.com/torydiary/think_tanks_and_campaigners/.

ConservativeHome. (2010). *Philip Booth: Why I have reservations about the "Nudge" philosophy*. Accessed 12 March 2016. http://www.conservativehome.com/platform/2010/03/philip-booth-why-i-have-reservations-about-the-nudge-philosophy.html.

ConservativeHome. (2011). *Andrew Lilico: Stick to the course, Mr Osborne. Don't let Ed Balls bore you into submission*. Accessed 21 April 2016. http://www.conservativehome.com/platform/2011/06/yaaawwn.html.

ConservativeHome. (2013). *Andrew Lilico: What Jonathan Portes gets wrong and why*. Accessed 15 October 2015. http://www.conservativehome.com/platform/2013/01/andrew-lilico-what-jonathan-portes-gets-wrong-and-why.html.

Converse, P. (2006 [1964]). The nature of belief systems in mass publics. *Critical Review: A Journal of Politics and Society, 18*(1), 1–74.

Cooren, F. (2016). *Organizational discourse: Communication and constitution*. London: Wiley.

Couldry, N. (2003). Media meta-capital: Extending the range of Bourdieu's field theory. *Theory & Society, 32*(5/6), 653–677.

Crehan, K. (2011). Gramsci's concept of common sense: A useful concept for anthropologists? *Journal of Modern Italian Studies, 16*(2), 273–287.

Crehan, K. (2016). *Gramsci's common sense: Inequality and its narratives*. Durham: Duke University Press.

Cronin, B. (2010). The diffusion of heterodox economics. *American Journal of Economics and Sociology, 69*(5), 1475–1494.

Crosthwaite, P. (2012). Is a financial crisis a trauma? *Cultural Critique, 82*(3), 34–67.

Crouch, C. (2011). *The strange non-death of neoliberalism*. Cambridge: Polity.

CUSPE. (2013). *Stewart Wallis—The flawed dominance of economics*. Accessed 4 November 2014. https://www.youtube.com/watch?v=7TbpQlcWaYQ.

Czarniawska, B. (1998). *A narrative approach to organization studies*. London: Sage.

Daily Telegraph. (2001). Norman still selling Portillo's dream. Accessed 15 March 2016. http://www.telegraph.co.uk/news/uknews/1334767/Norman-still-selling-Portillos-dream.html.

Daily Telegraph. (2009a). Free-market network demands bail-out for pharmaceutical industry. Accessed 20 February 2015. http://blogs.telegraph.co.uk/news/alexsingleton/8145947/freemarket_network_demands_bailout_for_pharmaceutical_industry/.

Daily Telegraph. (2009b). If David Cameron wants to govern, he should stop being afraid of ideas. Accessed 24 April 2016. http://www.telegraph.co.uk/comment/columnists/simonheffer/5700830/If-David-Cameron-wants-to-govern-he-should-stop-being-afraid-of-ideas.html.

Daily Telegraph. (2009c). Top 100 most influential Right-wingers, 50–51. Accessed 20 March 2016. http://www.telegraph.co.uk/news/politics/conservative/6256390/Top-100-most-influential-Right-wingers-50-1.html.

Daily Telegraph. (2009d). Madsen Pirie: The Archbishop of Canterbury caricatures consumers and fires at token targets. Accessed 18 February 2015. http://www.telegraph.co.uk/comment/personal-view/6315892/The-Archbishop-of-Canterbury-caricatures-consumers-and-fires-at-token-targets.html.

Daily Telegraph. (2009e). Madsen Pirie: Lord Stern is wrong: Giving up meat is no way to save the planet. Accessed 20 February 2015. http://www.telegraph.co.uk/news/earth/environment/climatechange/6445930/Lord-Stern-is-wrong-giving-up-meat-is-no-way-to-save-the-planet.html.

Daily Telegraph. (2009f). Sir Bryan Hopkin. Accessed 20 March 2016. http://www.telegraph.co.uk/news/obituaries/politics-obituaries/6638037/Sir-Bryan-Hopkin.html.

Daily Telegraph. (2010). Capital gains tax: Q&A. Accessed 20 March 2015. http://www.telegraph.co.uk/finance/personalfinance/capital-gains-tax/7771490/Capital-Gains-Tax-QandA.html.

Daily Telegraph. (2011). Neil O'Brien: Why the deficit deniers are deliberately missing the point. Accessed 15 March 2016. https://web.archive.org/web/20160326235251/http://blogs.telegraph.co.uk/news/neilobrien1/100074386/why-the-deficit-deniers-are-deliberately-missing-the-point/.

Daily Telegraph. (2012). It doesn't matter who funds think tanks, but if it did, left-wing ones would do particularly badly. Accessed 20 February 2015. http://blogs.telegraph.co.uk/finance/timworstall/100018107/it-doesnt-matter-who-funds-think-tanks-but-if-it-did-left-wing-ones-would-do-particularly-badly/.

Davenport, C. (2010). *Media bias, perspective, and state repression: The Black Panther Party*. Cambridge: Cambridge University Press.

Davies, W., & McGoey, L. (2012). Rationalities of ignorance: On financial crisis and the ambivalence of neoliberal epistemology. *Economy & Society, 41*(1), 64–83.

Dean, M. (2012). Rethinking neoliberalism. *Journal of Sociology, 50*(2), 150–163.

de Goede, M. (2009). Finance and the excess: The politics of visibility in the credit crisis. *Zeitschrift für Internationale Beziehungen, 16*(2), 295–306.

Denham, A., & Garnett, M. (1996). The nature and impact of think tanks in contemporary Britain. *Contemporary British History, 10*(1), 43–61.

Denham, A., & Garnett, M. (1998). *British think tanks and the climate of opinion*. London: UCL Press.

Denham, A., & Garnett, M. (2004). A 'hollowed out' tradition? British think tanks in the twenty-first century. In A. Denham & D. Stone (Eds.), *Think tank traditions* (pp. 232–246). Manchester: Manchester University Press.

Denham, A., & Garnett, M. (2006). What works? British think tanks and the end of ideology. *Political Quarterly, 77*(2), 156–165.

Denham, A., & Stone, D. (Eds.). (2004). *Think tank traditions*. Manchester: Manchester University Press.

Desai, R. (1994). Second hand dealers in ideas: Think tanks and Thatcherite hegemony. *New Left Review, 203*(1), 27–64.

Dewey, J. (1946 [1927]). *The public and its problems: An essay in political inquiry*. Chicago: Gateway Books.

Dexter, L. (2006). *Elite and specialized interviewing*. Colchester: ECPR Press.

DiMaggio, P., & Powell, W. (Eds.). (1991). *The new institutionalism in organizational analysis*. Chicago: University of Chicago Press.

Djelic, M. L. (2014). *Spreading ideas to change the world: Inventing and institutionalizing the neoliberal think tank* (Social Science Research Network). Accessed 22 February 2015. http://papers.ssrn.com/sol3/papers.cfm?abstract_id=2492010.

Dobry, M. (2009). *Sociologie des crises politiques*. Paris: Science Po, Les Presses.

Domhoff, G. W. (1967). *Who rules America?* New York: McGraw Hill.

Durkheim, E. (1995 [1912]). *The elementary forms of religious life*. New York: The Free Press.

DWP. (2010). *Consultation responses to 21st century welfare*. Accessed 25 March 2015. https://www.gov.uk/government/uploads/system/uploads/attachment_data/file/181144/21st-century-welfare-response.pdf.

Eisenhardt, K. (2002). Building theories from case study research. In M. Huberman & M. Miles (Eds.), *The qualitative researcher's companion* (pp. 4–36). Thousand Oaks: Sage.

Ekins, P. (1986). *The living economy: A new economics in the making*. London: Routledge.

Emirbayer, M., & Johnson, V. (2008). Bourdieu and organizational analysis. *Theory & Society, 37*, 1–44.

Engelen, E., Erturk, I., Froud, J., Johal, S., Leaver, A., Moran, M., et al. (2011). *Misrule of experts? The financial crisis as elite debacle* (CRESC Working Paper Series, 94).

European Students for Liberty. (2014). *Madsen Pirie: Prospects for liberty*. Accessed 25 March 2015. https://www.youtube.com/watch?v=BHP0GrBhuNs.

Evans, R. (1999). *Macroeconomic forecasting: A sociological appraisal*. London: Routledge.

Evans, K. (2011). 'Big society' in the UK: A policy review. *Children & Society, 25*, 164–171.

Eyal, G. (2013). Spaces between fields. In P. Gorski (Ed.), *Bourdieu and historical analysis* (pp. 158–182). Durham: Duke University Press.

Eyal, G., & Buchholz, L. (2010). From the sociology of intellectuals to the sociology of interventions. *Annual Review of Sociology, 36*, 117–137.

Eyal, G., & Levy, M. (2013). Economic indicators as public interventions. In T. Mata & S. Medema (Eds.), *The economist as public intellectual* (pp. 220–253). London: Duke University Press.

Eyal, G., & Pok, G. (2011). *From a sociology of professions to a sociology of expertise*. Accessed 20 February 2015. http://cast.ku.dk/papers_security_expertise/Eyal__2011_From_a_sociology_of_professions_to_a_sociology_of_expertise.pdf.

Eyerman, R. (2011). Intellectuals and cultural trauma. *European Journal of Social Theory, 14*(4), 453–467.

Fereday, J., & Muir-Cochrane, E. (2006). Demonstrating rigor using thematic analysis: A hybrid approach of inductive and deductive coding and theme development. *International Journal of Qualitative Methods, 5*(1), 1–11.

Festinger, L., Riecken, H., & Schachter, S. (2008 [1956]). *When prophecy fails: A social and psychological study of a modern group that predicted the destruction of the world*. London: Pinter & Martin.

Financial Times. (2008a). Think tank feels pinch as rival cashes in. Accessed 27 February 2016. http://www.ft.com/cms/s/0/a8990ad0-7453-11dd-bc91-0000779fd18c.html.

Financial Times. (2008b). Policy Exchange powers party's 'liberal revolution.' Accessed 2 March 2016. http://www.ft.com/cms/s/0/3006a6c4-26d1-11dd-9c95-000077b07658.html.

Financial Times. (2009a). Think tank says cuts will not pay debt. Accessed 15 October 2015. http://www.ft.com/cms/s/0/39c54894-7655-11de-9e59-00144feabdc0.html.

Financial Times. (2009b). Meet the new Tory establishment. Accessed 5 March 2016. http://www.ft.com/cms/s/2/ac5f0298-af38-11de-ba1c-00144feabdc0.html.

Financial Times. (2009c). Top bankers destroy value, study claims. Accessed 22 March 2016. https://www.ft.com/content/7e3edf6e-e827-11de-8a02-00144feab49a.

Financial Times. (2010a). UK out of recession, says think tank. Accessed 23 June 2015. http://www.ft.com/cms/s/0/271f57e6-0054-11df-8626-00144feabdc0.html.

Financial Times. (2010b). *Labour 'errors' attacked by former official*. Accessed 15 October 2015. http://www.ft.com/cms/s/0/d86bed74-4f38-11df-b8f4-00144feab49a.html.

Financial Times. (2010c). *Spending cuts spark fears for growth*. Accessed 15 October 2015. http://www.ft.com/cms/s/0/d7abbb20-7d65-11df-a0f5-00144feabdc0.html.

Financial Times. (2010d). *UK growing near long-term trend, says NIESR*. Accessed 30 June 2015. http://www.ft.com/cms/s/0/073c19f8-7590-11df-86c4-00144feabdc0.html.

Fischer, F., & Gottweis, H. (2012). *The argumentative turn revisited: Public policy as communicative practice*. London: Duke University Press.

Fischer, K., & Plehwe, D. (2013). Redes de think tanks e intelectuales de derecha en América Latina. *Nueva Sociedad, 245*, 70–86.

Fleck, L. (1979 [1935]). *The genesis and development of a scientific fact*. Chicago: University of Chicago Press.

Foucault, M. (1980). *Power, knowledge: Selected interviews and other writings*. New York: The Harvester Press.

Fourcade, M. (2009). *Economists and societies: Discipline and profession in the United States, Britain and France, 1890s to 1990s*. Princeton: Princeton University Press.

Frickel, S., & Gross, N. (2005). A general theory of scientific/intellectual movements. *American Sociological Review, 70*(2), 204–232.

Friedmann, J. (1992). *Empowerment: The politics of alternative development*. Hoboken: Wiley-Blackwell.

Friedman, J. (1997). What's wrong with libertarianism. *Critical Review, 11*(3), 403–467.

Frowen, S. (Ed.). (1983). *Controlling industrial economies: Essays in honour of Christopher Thomas Saunders*. London: Macmillan.

Gamble, A. (2009). *The spectre at the feast: Capitalist crisis and the politics of recession*. Basingstoke: Palgrave Macmillan.

Geertz, C. (1977). *The interpretation of cultures*. New York: Basic Books.

Gerring, J. (2007). *Case study research: Principles and practices*. Cambridge: Cambridge University Press.

Gibbons, M., Limoges, C., Nowotny, H., Schwartzman, S., Scott, P., & Trow, M. (1994). *The new production of knowledge: The dynamics of science and research in contemporary societies*. London: Sage.

Gieryn, T. (1995). Boundaries of science. In S. Jasanoff, G. Markle, J. Peterson, & T. Pinch (Eds.), *Handbook of science and technology studies* (pp. 393–443). Thousand Oaks, CA: Sage.

Gills, B. (2010). The return of crisis in the era of globalization: One crisis, or many? *Globalizations, 7*(1–2), 3–8.

Glaser, B., & Strauss, A. (1967). *The discovery of grounded theory: Strategies for qualitative research*. Chicago: Aldine.

González Hernando, M. (2018). Two British think tanks after the global financial crisis: intellectual and institutional transformations. *Policy & Society, 37*(2), 140–154.

González Hernando, M., & Baert, P. (forthcoming). Collectives of intellectuals: Their cohesiveness, accountability, and who can speak on their behalf. *The Sociological Review*.

González Hernando, M., Pautz, H., & Stone, D. (2018). Think tanks in 'hard times': The global financial crisis and economic advice. *Policy & Society, 37*(2), 125–139.

González Hernando, M., & Williams, K. (2018). Examining the link between funding regimes and intellectual interventions in universities and think tanks: A theoretical framework. *International Journal of Politics, Culture and Society, 31*(2), 193–206.

Gorski, P. (Ed.). (2013). *Bourdieu and historical analysis*. London: Duke University Press.

Gov.uk. (2010a). *Immigration: Home Secretary's speech of 5 November 2010*. Accessed 20 February 2016. https://www.gov.uk/government/speeches/immigration-home-secretarys-speech-of-5-november-2010.

Gov.uk. (2010b). *Nick Herbert's speech to the Policy Exchange*. Accessed 25 February 2016. https://www.gov.uk/government/speeches/nick-herberts-speech-to-the-policy-exchange.

Gov.uk. (2010c). *David Lidington: The Single Market is "essential to this government's agenda for trade and competitiveness."* Accessed 1 March 2016. https://www.gov.uk/government/news/the-single-market-is-essential-to-this-government-s-agenda-for-trade-and-competitiveness.

Gov.uk. (2010d). *Greg Clark: Three actions needed to help the Big Society grow.* Accessed 3 March 2016. https://www.gov.uk/government/news/three-actions-needed-to-help-the-big-society-grow.

Gov.uk. (2010e). *General Richards: Trading the perfect for the acceptable.* Accessed 3 March 2016. https://www.gov.uk/government/news/cds-on-the-sdsr-trading-the-perfect-for-the-acceptable.

Gov.uk. (2011). *Michael Gove on public sector pension reforms.* Accessed 3 March 2016. https://www.gov.uk/government/speeches/michael-gove-on-public-sector-pension-reforms.

Gov.uk. (2012a). *Damian Green: Making immigration work for Britain.* Accessed 28 February 2016. https://www.gov.uk/government/speeches/damian-greens-speech-on-making-immigration-work-for-britain.

Gov.uk. (2012b). *David Gauke: Where next for tackling tax avoidance?* Accessed 20 February 2016. https://www.gov.uk/government/speeches/speech-by-exchequer-secretary-to-the-treasury-david-gauke-mp-where-next-for-tackling-tax-avoidance.

Gov.uk. (2012c). *David Willetts: Our hi-tech future.* Accessed 3 March 2016. http://webarchive.nationalarchives.gov.uk/20121212135622/http://www.bis.gov.uk//news/speeches/david-willetts-policy-exchange-britain-best-place-science-2012.

Gov.uk. (2012d). *Alan Milburn and Neil O'Brien set to lead the drive to improve social mobility and reduce child poverty.* Accessed 25 April 2016. https://www.gov.uk/government/news/alan-milburn-and-neil-obrien-set-to-lead-the-drive-to-improve-social-mobility-and-reduce-child-poverty.

Gov.uk. (2013a). *Lord Freud: Improving employment outcomes.* Accessed 20 February 2016. https://www.gov.uk/government/speeches/improving-employment-outcomes–2.

Gov.uk. (2013b). *Michael Fallon: Royal Mail—Ensuring long term success.* Accessed 1 March 2016. https://www.gov.uk/government/speeches/royal-mail-ensuring-long-term-success.

Gov.uk. (2013c). *Francis Maude: Ministers and mandarins—Speaking truth unto power.* Accessed 1 March 2016. https://www.gov.uk/government/speeches/ministers-and-mandarins-speaking-truth-unto-power.

Gov.uk. (2013d). *Government announces new member to the Social Security Advisory Committee.* Accessed 25 April 2016. https://www.gov.uk/government/news/government-announces-new-member-to-the-social-security-advisory-committee.

Gramsci, A. (1999 [1971]). *Selections from the prison notebooks.* London: Elecbooks.

Greenham, T., Jackson, A., Ryan-Collins, J., & Werner, R. (2012). *Where does money come from: A guide to the UK monetary and banking system.* London: NEF.

Guardian. (2005). Full text: David Cameron's speech to Policy Exchange. Accessed 15 March 2016. http://www.theguardian.com/politics/2005/jun/29/speeches.conservatives.

Guardian. (2008a). *Andrew Simms: Tackling the 'triple crunch' with a green new deal.* Accessed 28 October 2014. https://www.theguardian.com/commentisfree/2008/sep/19/creditcrunch.marketturmoil.

Guardian. (2008b). Andrew Simms: The final countdown. Accessed 28 October 2014. https://www.theguardian.com/environment/2008/aug/01/climatechange.carbonemissions.

Guardian. (2008c). Will the market bounce back soon? Accessed 16 February 2015. http://www.theguardian.com/money/2008/nov/01/moneyinvestments-investmentfunds?INTCMP=SRCH.

Guardian. (2008d). 167 Eamonn Butler: Sentamu and the City. Accessed 18 February 2015. http://www.theguardian.com/commentisfree/2008/sep/25/anglicanism.religion.

Guardian. (2008e). *No recession, economists say—Provided there are no more rocks.* Accessed 20 October 2015. http://www.theguardian.com/business/2008/may/02/creditcrunch.banking.

Guardian. (2008f). *Cameron condemns 'insane' report on northern cities.* Accessed 15 March 2016. http://www.theguardian.com/politics/2008/aug/13/davidcameron.conservatives1.

Guardian. (2008g). *What can they be thinking?* Accessed 28 February 2016. http://www.theguardian.com/politics/2008/sep/26/thinktanks.conservatives.

Guardian. (2009). Revealed: £3bn mistake in George Osborne's budget plan. Accessed 21 October 2015. http://www.theguardian.com/politics/2009/oct/09/george-osborne-budget-deficit.

Guardian. (2010a). Anna Coote: A shorter working week would benefit society. Accessed 30 October 2014. http://www.theguardian.com/commentisfree/2010/jul/30/short-working-week-benfit-society.

Guardian. (2010b). Eamonn Butler: Britain's bloated government needs to go on a diet. Accessed 20 February 2015. http://www.theguardian.com/commentisfree/2010/jun/08/britains-bloated-government.

Guardian. (2010c). Spending cuts 'will see rise in absolute child poverty.' Accessed 28 March 2016. http://www.theguardian.com/politics/2010/dec/16/spending-cuts-rise-absolute-child-poverty.

Guardian. (2012a). James Meadway: Development's fat cats have been gorging on private sector values. Accessed 20 March 2015. http://www.theguardian.com/commentisfree/2012/sep/19/development-fat-cats-private-sector-values.

Guardian. (2012b). Compassionate conservatives find it's time to think again. Accessed 28 March 2016. http://www.theguardian.com/politics/2012/mar/05/compassionate-conservatives.

Guardian. (2013). Romanian and Bulgarian migrants 'unlikely to head for UK'—Report. Accessed 25 October 2015. https://www.theguardian.com/uk/2013/apr/05/eastern-european-migrants-report.

Guardian. (2014). Stewart Wallis: An economic system that supports people and planet is still possible. Accessed 4 November 2014. http://www.theguardian.com/sustainable-business/2014/nov/04/economic-system-supports-people-planet-possible.

Guy Peters, B. (2012). *Institutional theory in political science*. London: Continuum.

Haas, E. (1989). Do regimes matter? Epistemic communities and evolving policies to control Mediterranean pollution. *International Organisation, 43*(3), 377–403.

Hajer, M. (1993). Discourse coalitions: The case of acid rain in Great Britain. In F. Fischer & J. Forester (Eds.), *The argumentative turn in policy planning* (pp. 43–76). London: University College Press.

Hall, P. A. (1993). Policy paradigms, social learning, and the state: The case of economic policymaking in Britain. *Comparative Politics, 25*(3), 275–296.

Hallett, T. (2003). Symbolic power and organizational culture. *Sociological Theory, 21*(2), 128–149.

Harré, R., Moghaddam, F. M., Pilkerton Cairnie, T., Rothbart, D., & Sabat, S. R. (2009). Recent advances in positioning theory. *Theory and Psychology, 19*(5), 5–31.

Harvey, W. (2011). Strategies for conducting elite interviews. *Qualitative Research, 11*(4), 431–441.

Hay, C. (2011). Ideas and the construction of interest. In D. Béland & R. Cox (Eds.), *Ideas and politics in social science research* (pp. 65–82). New York: Oxford University Press.

Hayton, R. (2012). *Reconstructing conservatism? The Conservative party in opposition, 1997–2010*. Manchester: Manchester University Press.

Heffernan, R. (1996). Blueprint for a revolution? The politics of the Adam Smith Institute. *Contemporary British History, 10*(1), 73–87.

Herbst, S. (2003). Political authority in a mediated age. *Theory & Society, 32*(4), 481–503.

Hierländer, R., Huber, P., Iara, A., Landesmann, M., Nowotny, K., O'Mahony, M., et al. (2010). *Migration, skills and productivity* (WIIW Research Reports, 365). Accessed 20 June 2015. http://wiiw.ac.at/migration-skills-and-productivity-dlp-2270.pdf.

Hilgartner, S. (2000). *Science on stage: Expert advice as public drama*. Stanford: Stanford University Press.

HM Government. (2012). *Civil service reform plan*. Accessed 20 February 2016. https://www.gov.uk/government/publications/civil-service-reform-plan.

HM Government. (2013). *Civil service reform plan: One year on*. Accessed 20 February 2016. https://www.gov.uk/government/publications/civil-service-reform-plan-one-year-on–2.

Holmwood, J. (2013). Rethinking moderation in a pragmatist frame. *The Sociological Review, 61*(2), 180–195.

Holmwood, J. (2014). Sociology's past and futures: The impact of external structure, policy and financing. In J. Holmwood & J. Scott (Eds.), *A handbook of British sociology*. London: Palgrave Macmillan.

Holmwood, J., Smith, T., & Thomas, A. (2013). Sociologies of moderation. *The Sociological Review, 61*(2), 6–17.

Independent. (1998). Obituary: Christopher Dow. Accessed 20 March 2016. http://www.independent.co.uk/arts-entertainment/obituary-christopher-dow-1189026.html.

Independent. (2008). Tories' favourite think tank sued by Muslim group. Accessed 18 May 2016. http://www.independent.co.uk/news/uk/politics/tories-favourite-think-tank-sued-by-muslim-group-897548.html.

International Business Times. (2014). Rise of the new Libertarians: Meet Britain's next political generation. Accessed 18 February 2015. http://www.ibtimes.co.uk/rise-new-libertarians-meet-britains-next-political-generation-1469233.

Jackson, B. (2012). The think tank archipelago: Thatcherism and neoliberalism. In B. Jackson & R. Saunders (Eds.), *Making Thatcher's Britain* (pp. 43–61). Cambridge: Cambridge University Press.

Jackson, T. (2011 [2009]). *Prosperity without growth: Economics for a finite planet*. London: Routledge.

Jacobs, R., & Townsley, E. (2011). *The space of opinion: Media intellectuals and the public sphere*. Oxford: Oxford University Press.

Jacoby, R. (2000). *The last intellectuals: American culture in the age of academe*. New York: Basic Books.

Jacques, P., Dunlap, R., & Freeman, M. (2008). The organisation of denial: Conservative think tanks and environmental scepticism. *Environmental Politics, 17*(3), 349–385.

James, S. (1993). The idea brokers: The impact of think tanks on British government. *Public Administration, 71*, 491–506.

Jasanoff, S. (1995). *Science at the bar: Law, science, and technology in America*. Cambridge, MA: Harvard University Press.

Jones, K. (1994). *An economist among mandarins: A biography of Robert Hall*. Cambridge: Cambridge University Press.

Kay, L., Smith, K., & Torres, J. (2013). Think tanks as research mediators? Case studies from public health. *Evidence and Policy, 59*(3), 371–390.

Kerr, R., & Robinson, S. (2009). The hysteresis effect as creative adaptation of the habitus: Dissent and transition to the 'corporate' in post-Soviet Ukraine. *Organization, 16*(6), 829–853.

Kingdon, J. (2003). *Agendas, alternatives and public policies*. New York: Longman.

Klamer, A. (2007). *Speaking of economics: How to get in the conversation*. London: Routledge.

Klamer, A., McCloskey, R., & Solow, R. (Eds.). (1988). *The consequences of economic rhetoric*. Cambridge: Cambridge University Press.

Koselleck, R. (2002). *The practice of conceptual history*. Stanford: Stanford University Press.

Krastev, I. (2001). Think tanks: Making and faking influence. *Southeast European and Black Sea Studies, 1*(2), 17–38.

Ladi, S. (2011). Think tanks, discursive institutionalism and policy change. In G. Papanagnou (Ed.), *Social science and policy challenges: Democracy, values and capacities*. Paris: UNESCO.

Lakoff, G. (2004). *Don't think of an elephant! Know your values and frame the debate*. London: Chelsea Green.

Lamb, B. (2014). Is charity campaigning under threat from the coalition government? *Voluntary Sector Review, 5*(1), 125–138.

Lawson, T. (2009). The current economic crisis: Its nature and the course of academic economics. *Cambridge Journal of Economics, 33*(4), 759–777.

Libertarian Home. (2014). *Sam Bowman promoted to deputy director of the Adam Smith Institute*. Accessed 27 March 2015. http://libertarianhome.co.uk/2014/11/sam-bowman-promoted-to-deputy-director-of-the-adam-smith-institute/.

List, C., & Pettit, P. (2011). *Group agency: The possibility, design, and status of corporate groups*. Oxford: Oxford University Press.

Lo, A. (2011). *Reading about the financial crisis: A 21-book review*. Social Science Research Network. Accessed 15 March 2013. http://ssrn.com/abstract=1949908.

London Review of Books. (2013). *Jonathan Portes: An exercise in scapegoating*. Accessed 10 October 2015. http://www.lrb.co.uk/v35/n12/jonathan-portes/an-exercise-in-scapegoating.

Lilico, A. (2009). *What killed capitalism? The crisis: What caused it and how to respond*. London: Centre for Policy Studies.

Lischinsky, A. (2011). In times of crisis: A corpus approach to the construction of the global financial crisis in annual reports. *Critical Discourse Studies, 8*(3), 153–168.

LSE. (2011). *Keynes v Hayek*. Accessed 20 March 2015. http://www.lse.ac.uk/newsAndMedia/videoAndAudio/channels/publicLecturesAndEvents/player.aspx?id=1107.

LSE. (2012). *About time: Examining the case for a shorter working week*. Accessed 21 April 2015. https://www.youtube.com/watch?v=nqI951u9emQ.

Lucas, R. (2003). Macroeconomic priorities. *American Economic Review, 93*(1), 1–14.

Markoff, J., & Montecinos, V. (1993). The ubiquitous rise of economists. *Journal of Public Policy, 13*(1), 37–68.

Mazur, A. (1998). *A hazardous inquiry: The Rashomon effect at Love Canal*. Cambridge, MA: Harvard University Press.

McDowell, L. (1998). Elites in the city of London: Some methodological considerations. *Environment and Planning A, 30*(12), 2133–2146.

McGann, J. (2009). *2008 global go to think tanks and policy advice ranking*. Think Tanks and Civil Society Program. University of Pennsylvania.

McGann, J. (2010a). *2009 global go to think tanks and policy advice ranking*. Think Tanks and Civil Society Program. University of Pennsylvania.

McGann, J. (2010b). *Democratization and market reform in developing and transitional countries: Think tanks as catalysts*. London: Routledge.

McGann, J. (2011). *2010 global go to think tanks and policy advice ranking*. Think Tanks and Civil Society Program. University of Pennsylvania.
McGann, J. (2012). *2011 global go to think tanks and policy advice ranking*. Think Tanks and Civil Society Program. University of Pennsylvania.
McGann, J. (2013). *2012 global go to think tanks and policy advice ranking*. Think Tanks and Civil Society Program. University of Pennsylvania.
McGann, J. (2014). *2013 global go to think tanks and policy advice ranking*. Think Tanks and Civil Society Program. University of Pennsylvania.
McGann, J. (2015). *2014 global go to think tanks and policy advice ranking*. Think Tanks and Civil Society Program. University of Pennsylvania.
McGann, J. (2016). *2014 global go to think tanks and policy advice ranking*. Think Tanks and Civil Society Program. University of Pennsylvania.
McGann, J. (2017). *2016 global go to think tanks and policy advice ranking*. Think Tanks and Civil Society Program. University of Pennsylvania.
McGann, J. (2018). *2017 global go to think tanks and policy advice ranking*. Think Tanks and Civil Society Program. University of Pennsylvania.
McGann, J., Viden, A., & Rafferty, J. (Eds.). (2014). *How think tanks shape social development policies*. Philadelphia: University of Pennsylvania Press.
McGoey, L. (2012). Strategic unknowns: Towards a sociology of ignorance. *Economy & Society, 41*(1), 1–16.
McLennan, G. (2004a). *Dynamics of transformative ideas in contemporary public discourse, 2002–2003*. Accessed 15 October 2013. http://www.esds.ac.uk/doc/5312/mrdoc/pdf/q5312uguide.pdf.
McLennan, G. (2004b). Travelling with vehicular ideas: The case of the third way. *Economy & Society, 33*(4), 484–499.
McNutt, K., & Marchildon, G. (2009). Think tanks and the web: Measuring visibility and influence. *Canadian Public Policy, 35*(2), 219–236.
Meadows, D. H., Meadows, D. L., Randers, J., & Behrens III, W. W. (1972) *The limits to growth: A report for the club of Rome's project on the predicament of mankind*. New York: Universe Books.
Medium. (2015). *Ed Conway: Why Jonathan Portes left NIESR*. Accessed 30 October 2015. https://medium.com/@edconwaysky/why-jonathan-portes-left-niesr-3999329290d9#.t9pn5q6ka.
Medvetz, T. (2006). The strength of weekly ties: Relations of material and symbolic exchange in the conservative movement. *Politics & Society, 34*(3), 343–368.
Medvetz, T. (2008). *Think tanks as an emergent field*. New York: Social Science Research Council. Accessed 4 April 2014. http://www.ssrc.org/publications/view/A2A2BA10-B135-DE11-AFAC-001CC477EC70/.

Medvetz, T. (2012a). *Think tanks in America*. Chicago: University of Chicago Press.

Medvetz, T. (2012b). Murky power: 'Think tanks' as boundary organizations. In D. Golsorkhi, D. Courpasson, & J. Sallaz (Eds.), *Rethinking power in organizations, institutions, and markets: Research in the sociology of organizations* (pp. 113–133). Bingley: Emerald Group Publishing.

Medvetz, T. (2012c). Scholar as 'sitting duck': The Cronon affair and the buffer zone in American public debate. *Public Culture, 24*(1), 47–53.

Medvetz, T. (2013). Opening the black box of intellectual autonomy. *Papers: Revista de Sociología, 98*(3), 573–579.

Medvetz, T. (2015). Field theory and organisational power: Four modes of influence among policy 'think tanks'. In M. Hilgers & E. Mangez (Eds.), *Bourdieu's theory of social fields*. London: Routledge.

Mikecz, R. (2012). Interviewing elites: Addressing methodological issues. *Qualitative Inquiry, 18*(6), 482–493.

Mills, C. W. (1956). *The power elite*. Oxford: Oxford University Press.

Mirowski, P. (2002). *Machine dreams: Economics becomes a cyborg science*. Cambridge: Cambridge University Press.

Mirowski, P. (2013). *Never let a serious crisis go to waste*. London: Verso.

Mirowski, P., & Plehwe, D. (2009). *The road to Mont Pèlerin: The making of the neoliberal thought collective*. Cambridge, MA: Harvard University Press.

Misztal, B. (2007). *Intellectuals and the public good: Creativity and civil courage*. Cambridge: Cambridge University Press.

Misztal, B. (2012). Public intellectuals and think tanks: A free market in ideas? *International Journal of Politics, Culture and Society, 25*(4), 127–141.

Morgan, G. (2011). The use of UK charity accounts data for researching the performance of voluntary organisations. *Voluntary Sector Review, 2*(2), 213–230.

Morin, E. (1976). Pour une crisologie. *Communications, 25*(1), 149–163.

Morris, D. (2012). Charities and the big society: A doomed coalition? *Legal Studies, 32*(1), 132–153.

Muller, C. (1996). The institute of economic affairs: Undermining the post-war consensus. *Contemporary British History, 10*(1), 88–110.

Muniesa, F. (2014). *The provoked economy: Economic reality and the performative turn*. London: Routledge.

Natorski, M. (2015). Epistemic (un)certainty in times of crisis: The role of coherence as a social convention in the European Neighbourhood Policy after the Arab Spring. *European Journal of International Relations, 22*(3), 646–670.

New Statesman. (2011). Jonathan Portes: The coalition's confidence trick. Accessed 10 October 2015. http://www.newstatesman.com/economy/2011/08/interest-rates-debt-government.

Norman, J. (2010). *The big society: The anatomy of the new politics.* Buckingham: University of Buckingham Press.

Norman, J. (2012). *Fisking Portes: How politically neutral *is* the Director of NIESR?* Accessed 23 October 2015. http://www.jessenorman.com/2012/11/fisking-portes-how-politically-neutral-is-the-director-of-niesr.html.

Norpoth, H. (1991). The popularity of the Thatcher government: A matter of war and economy. In H. Norpoth, M. Lewis-Beck, & J. Lafay (Eds.), *Economics and politics: The calculus of support.* Ann Arbor: University of Michigan Press.

Office for National Statistics. (2014). *Gross Domestic Product preliminary estimate, Q1 2014.* Accessed 1 April 2016. http://www.ons.gov.uk/ons/rel/gva/gross-domestic-product–preliminary-estimate/q1-2014/stb-gdp-preliminary-estimate–q1-2014.html.

Oliver, M. (1996). A response to Denham and Garnett's 'the nature and impact of think tanks in contemporary Britain'. *Contemporary British History, 10*(2), 80–86.

Onthinktanks.org. (2013). *For-profit think tanks and implications for funders.* Accessed 25 March 2015. https://onthinktanks.org/articles/for-profit-think-tanks-and-implications-for-funders/.

Osborne, T. (2004). On mediators: Intellectuals and the ideas trade in the knowledge society. *Economy & Society, 33*(4), 430–447.

Pain, N., & Young, G. (2004). The macroeconomic impact of UK withdrawal from the EU. *Economic Modelling, 21,* 387–408.

ParliamentLive. (2012). *Treasury Committee: Tuesday 13 November 2012.* Accessed 10 October 2015. http://www.parliamentlive.tv/Event/Index/a31b4072-101e-4a65-8a90-f14cba88937d.

Pautz, H. (2010). Think tanks in the United Kingdom and Germany: Actors in the modernisation of social democracy. *British Journal of Politics and International Relations, 12*(2), 274–294.

Pautz, H. (2012a). *Think tanks, social democracy and social policy.* London: Palgrave Macmillan.

Pautz, H. (2012b). The think tanks behind 'cameronism'. *British Journal of Politics and International Relations, 15*(3), 362–377.

Pautz, H. (2016). Managing the crisis? Think tanks and the British response to global financial crisis and great recession. *Critical Policy Studies, 11*(2), 191–210 [Online early access].

Pels, D. (1995). Knowledge politics and anti-politics: Toward a critical appraisal of Bourdieu's concept of intellectual autonomy. *Theory & Society, 24*(1), 79–104.

Pettifor, A. (2003). *The real world economic outlook 2003: The legacy of globalization: Debt and deflation*. London: Palgrave Macmillan.

Pettifor, A. (2006). *The coming first world debt crisis*. London: Palgrave Macmillan.

Pickett, R., & Wilkinson, K. (2010). *The spirit level: Why more equal societies almost always do better*. London: Allen Lane.

Pielke, R. (2007). *The honest broker: Making sense of science in policy and politics*. Cambridge: Cambridge University Press.

Pieria. (2013). *Jonathan Portes: NIESR and the European Commission: A response to Dan Hannan*. Accessed 12 October 2015. http://www.pieria.co.uk/articles/niesr_and_the_european_commission_a_response_to_dan_hannan.

Pierson, R. (1994). The epistemic authority of expertise. In *PSA: Proceeding of the Biennial Meeting of the Philosophy of Science Association* (Vol. 1, pp. 398–405).

Pierson, P. (2004). *Politics in time: History, institutions and social analysis*. Princeton: Princeton University Press.

Pirie, I. (2012). Representations of economic crisis in contemporary Britain. *British Politics, 7*(4), 341–364.

Pirie, M. (1988). *Micropolitics: The creation of successful policy*. Aldershot: Wildwood House.

Pirie, M. (2007). *How to win every argument: The use and abuse of logic*. London: Continuum.

Pirie, M. (2008). *Freedom 101*. London: Adam Smith Institute.

Pirie, M. (2009). *Zero base policy*. London: Adam Smith Institute.

Pirie, M. (2012a). *Think tank: The history of the Adam Smith Institute*. London: Biteback.

Pirie, M. (2012b). *Economics made simple: How money, trade and markets really work*. Petersfield: Harriman House.

Plehwe, D. (2010). Think tanks und Entwicklung. Bessere Integration von Wissenschaft und Gesellschaft? *Journal für Entwicklungspolitik, 26*(2), 9–37.

Porter, T. (1995). *Trust in numbers: The pursuit of objectivity in science and public life*. Princeton, NJ: Princeton University Press.

Posner, R. (2003). *Public intellectuals: A study of decline*. Cambridge, MA: Harvard University Press.

Prospect. (2010). Think tank of the year awards—The winners. Accessed 20 February 2016. http://www.prospectmagazine.co.uk/politics/think-tank-of-the-year-awards-the-winners.

Prospect. (2013). The Prospect duel: Was Osborne right? Accessed 10 October 2015. http://www.prospectmagazine.co.uk/regulars/the-prospect-duel-mark-littlewood-jonathan-portes-was-osborne-right.

Rantanen, T. (2012). In nationalism we trust? In J. Caraça, G. Cardoso, & M. Castells (Eds.), *Aftermath: The cultures of economic crisis* (pp. 132–153). Oxford: Oxford University Press.

Reay, D. (2004). It's all becoming a habitus': Beyond the habitual use of Pierre Bourdieu's concept of habitus in educational research. *British Journal of Sociology of Education, 25*(4), 431–444.

Reay, D., Ball, S., & David, M. (2001). Making a difference?: Institutional habituses and higher education choice. *Sociological Research Online, 5*(4), 126–142.

Redden, J. (2010). *The mediation of poverty: The news, new media and politics* (PhD thesis). Department of Media and Communications, Goldsmiths, University of London. Accessed 5 June 2015. http://research.gold.ac.uk/6540/.

Reichmann, W. (2011). Institutionalizing scientific knowledge: The social and political foundation of empirical economic research. *Sociology Compass, 5*(7), 564–575.

Reinhart, C., & Sbrancia, M. B. (2011). *The liquidation of government debt* (NBER Working Papers, 16893). Accessed 12 March 2016. http://www.nber.org/papers/w16893.pdf.

Reuters. (2009). *On wealth versus well-being*. Accessed 20 October 2014. http://blogs.reuters.com/davos/2009/02/01/on-wealth-versus-well-being/.

Rich, A. (2011). Ideas, expertise and think tanks. In D. Béland & R. Cox (Eds.), *Ideas and politics in social science research* (pp. 191–208). New York: Oxford University Press.

Rich, A., & Weaver, K. (2000). Think tanks in the U.S. media. *The Harvard International Journal of Press/Politics, 5*(4), 81–103.

Ricoeur, P. (1980). Narrating time. *Critical Inquiry, 7*(1), 169–190.

Rohloff, A., & Wright, S. (2010). Moral panic and social theory: Beyond the heuristic. *Current Sociology, 58*(3), 403–419.

Roitman, J. (2013). *Anti-crisis*. London: Duke University Press.

Rosanvallon, P. (2008). *La légitimité démocratique: Impartialité, réflexivité, proximité*. Paris: Seuil.

Roth, W., & Mehta, J. (2002). The Rashomon effect: Combining positivist and interpretivist approaches in the analysis of contested events. *Sociological Methods and Research, 31*(2), 131–173.

Sabatier, P. (Ed.). (1999). *Theories of the policy process*. Boulder: Westview Press.

Saïd, E. (1994). *Representations of the intellectual*. New York: Vintage Books.

Sandvoss, C. (2010). Conceptualizing the global economic crisis in popular communication research. *Popular Communication: The International Journal of Media and Culture, 8*(3), 154–161.

Sapiro, G. (2009). Modèles d'intervention politique des intellectuels: Le cas français. *Actes de la recherche en sciences sociales, 176–177*(1–2), 8–31.

Savage, M., Warde, A., & Devine, F. (2005). Capitals, assets, and resources: Some critical issues. *British Journal of Sociology, 56*(1), 31–47.

Schlesinger, P. (2009). Creativity and the experts: New Labour, think tanks, and the policy process. *International Journal of Press/Politics, 14*(1), 3–20.

Schlichter, D. (2011). *Paper money collapse: The folly of elastic money and the coming monetary breakdown*. London: Wiley.

Schmidt, V. (2008). Discursive institutionalism: The explanatory power of ideas and discourse. *Political Science, 11*(1), 303–322.

Schmidt, V., & Thatcher, M. (2013). *Resilient liberalism in Europe's political economy*. Cambridge: Cambridge University Press.

Schumacher, D. (2011). *Small is beautiful in the XXI century: The legacy of E.F. Schumacher*. Devon: Green Books.

Schumacher, E. F. (1973). *Small is beautiful: Economics as if people mattered*. London: Blond & Briggs.

Scott, R. (1995). *Institutions and organizations*. Thousand Oaks: Sage.

Scott Solomon, M. (2010). Critical ideas in times of crisis: Reconsidering Smith, Marx, Keynes, and Hayek. *Globalizations, 7*(1–2), 127–135.

Scottish Liberty Forum. (2013). *Sam Bowman: The free market road to social justice*. Accessed 30 March 2016. https://www.youtube.com/watch?v=Rptm0yyvPik.

Seddon, N. (2009). *Who cares? How state funding and political activism change charity*. London: Civitas.

Seiler, C., & Wohlrabe, K. (2010). A critique of the 2009 global "go-to think tank" ranking. *CESifo DICE Report, 8*(2), 60–63.

Selgin, G. (1994). Are banking crises free-market phenomena? *Critical Review: A Journal of Politics and Society, 8*(4), 591–608.

Silva, P. (2009). *In the name of reason: Technocrats and politics in Chile*. University Park: Penn State University Press.

Simms, A. (2013). *Cancel the Apocalypse: A new path to prosperity*. London: Little, Brown.

Simpson, D. (2013). *The rediscovery of classical economics: Complexity, adaptation and growth*. Cheltenham: Edward Elgar Publishing.

Sinclair, M. (2009). *Taxpayer funded lobbying and political campaigning*. London: The TaxPayers' Alliance.

Sinclair, T. (2010). Round up the usual suspects: Blame and the subprime crisis. *New Political Economy, 15*(1), 91–107.

Skidelsky, R., Jackson, T., Schor, J., Coote, A., Franklin, J., Harris, M., et al. (2013). *Time on our side: Why we all need a shorter working week*. London: NEF.

Snowdon, C. (2012). *Sock puppets: How the government lobbies itself and why*. London: Institute of Economic Affairs.

Stahl, J. (2016). *Right moves: The conservative think tank in American political culture since 1945*. Chapel Hill: University of North Carolina Press.

Stephens, N. (2007). Collecting data from elites and ultra elites: Telephone and face-to-face interviews with macroeconomists. *Qualitative Research, 7*(2), 203–216.

Stone, D. (1991). Old guard versus new partisans: Think tanks in transition. *Australian Journal of Political Science, 26*(2), 197–215.

Stone, D. (1996a). From the margins of politics: The influence of think-tanks in Britain. *West European Politics, 19*(4), 675–692.

Stone, D. (1996b). *Capturing the political imagination: Think tanks and the policy process*. London: Frank Cass.

Stone, D. (2007). Recycling bins, garbage cans or think tanks? Three myths regarding policy analysis institutes. *Public Administration, 85*(2), 259–278.

Sunstein, C., & Thaler, R. (2008). *Nudge: Improving decisions about health, wealth, and happiness*. New Haven: Yale University Press.

Swartz, D. (2008). Bringing Bourdieu's master concepts into organizational analysis. *Theory & Society, 37*(1), 45–52.

Swartz, D. (2013). *Symbolic power, politics and intellectuals: The political sociology of Pierre Bourdieu*. Chicago: University of Chicago Press.

Tax Justice Network. (2010). *Adam Smith, the Adam Smith Institute, and flat taxes*. Accessed 16 February 2015. http://taxjustice.blogspot.co.uk/2010/01/adam-smith-adam-smith-institute-and.html.

Taylor, M. (2011). Think tanks, public policy and academia. *Public Money & Management, 31*(1), 10–11.

Tchilingirian, J. (2015). *British think tanks and the production of policy knowledge: A social network analysis of policy intellectuals* (PhD thesis). Department of Sociology, University of Cambridge.

TED. (2010). *Nic Marks: The Happy Planet Index*. Accessed 24 October 2014. http://www.ted.com/talks/nic_marks_the_happy_planet_index.

Tesseyman, A. J. (1999). *The new right think tanks and policy change in the UK* (PhD thesis). Department of Politics, University of York.

Thaa, W. (2012). Deliberating experts versus political representation. In S. Brooks, D. Stasiak, & T. Zyro (Eds.), *Policy expertise in contemporary democracies* (pp. 57–67). Farnham: Ashgate.

t'Hart, P., & Tindall, K. (Eds.). (2009). *Framing the global economic downturn: Crisis rhetoric and the politics of recessions*. Sydney: ANU Press.

The Observer. (2013). Health groups dismayed by news 'big tobacco' funded rightwing thinktanks. Accessed 20 February 2015. http://www.theguardian.com/society/2013/jun/01/thinktanks-big-tobacco-funds-smoking.

The Spectator. (2015a). *Niall Ferguson: Jonathan Portes, master of the political correction*. Accessed 10 October 2015. http://www.spectator.co.uk/2015/06/jonathan-portes-master-of-correct-politicalness/.

The Spectator. (2015b). *Jonathan Portes fired as NIESR director*. Accessed 30 October 2015. http://blogs.spectator.co.uk/2015/10/jonathan-portes-stands-down-from-niesr-by-mutual-consent/.

The Times. (2004). Brown attacked for fiscal 'fudge.' Accessed 24 October 2015. http://www.thetimes.co.uk/tto/business/entrepreneur/article2229114.ece.

Think Tank Watch. (2014). Are Prospect magazine's think tank of the year awards rigged? Accessed 27 October 2014. http://www.thinktankwatch.com/2014/07/is-prospect-magazines-think-tank-of.html.

Third Sector. (2016). *Anti-lobbying clause put on hold by government*. Accessed 25 May 2016. http://www.thirdsector.co.uk/anti-lobbying-clause-put-hold-government/policy-and-politics/article/1393042.

Thompson, J. (1995). *Books in the digital age: The transformation of academic and higher education publishing in Britain and the United States*. Cambridge: Polity.

Thompson, J. (2012). The metamorphosis of a crisis. In J. Caraça, G. Cardoso, & M. Castells (Eds.), *Aftermath: The cultures of economic crisis* (pp. 59–81). Oxford: Oxford University Press.

Tooze, A. (2019). *Crashed: How a decade of financial crises changed the world*. London: Penguin.

Transparify. (2016). *How transparent are think tanks about who funds them 2016? A survey of 200 think tanks in 47 countries worldwide*. Accessed 20 March 2016. http://static1.squarespace.com/static/52e1f399e4b06a94c0cdaa41/t/5773022de6f2e1ecf70b26d1/1467154992324/Transparify+2016+Think+Tanks+Report.pdf.

Trevisan, J. (2012). Global go-to 2011 think tank ranking: An analysis. *International Center for Climate Governance*. Accessed 25 February 2015. http://www.iccgov.org/FilePagineStatiche/Files/Publications/Reflections/03_reflection_february_2012.pdf.

Van Langenhove, L., & Harré, R. (1999). Introducing positioning theory. In R. Harré & L. Van Langenhove (Eds.), *Positioning theory: Moral contexts of international action*. Oxford: Wiley-Blackwell.

Viveiros de Castro, E. (2014). *Who is afraid of the ontological wolf? Some comments on an ongoing anthropological debate*. CUSAS Annual Marilyn Strathern Lecture, Cambridge. Accessed 20 November 2015. https://www.academia.edu/12865685/Who_is_afraid_of_the_ontological_wolf.

Wacquant, L. (1989). For a socio-analysis of intellectuals: On "homo academicus". *Berkeley Journal of Sociology, 34*, 1–29.

Wade, R. (2013). *Conservative Party economic policy: From Heath in opposition to Cameron in coalition*. Basingstoke: Palgrave Macmillan.

Walby, S. (2016). *Crisis*. Cambridge: Polity.

Washington Post. (2015). How Elizabeth Warren picked a fight with Brookings and won. Accessed 25 October 2015. https://www.washingtonpost.com/politics/how-elizabeth-warren-picked-a-fight-with-brookings–and-won/2015/09/29/bfe45276-66c7-11e5-9ef3-fde182507eac_story.html.

Washington Post. (2017). Trump could cause 'the death of think tanks as we know them.' Accessed 12 May 2017. https://www.washingtonpost.com/opinions/global-opinions/trump-could-cause-the-death-of-think-tanks-as-we-know-them/2017/01/15/8ec3734e-d9c5-11e6-9a36-1d296534b31e_story.html?utm_term=.48b548c7a5f9.

Weale, M. (2010). Unsustainable consumption: The structural flaw behind the UK's long boom. In G. Giudice, R. Kuenzel, & T. Springbett (Eds.), *UK Economy: The crisis in perspective* (pp. 59–79). London: Routledge.

Weaver, R. K. (1989). The changing world of think tanks. *PS: Political Science and Politics, 22*(3), 563–578.

Weingart, P. (1999). Scientific expertise and political accountability: Paradoxes of science in politics. *Science and Public Policy, 26*(3), 151–161.

Weir, M. (1992). Ideas and the politics of bounded innovation. In S. Steinmo, K. Thelen, & F. Longstreth (Eds.), *Historical institutionalism in comparative analysis*. Cambridge: Cambridge University Press.

Westermeier, C. (2018). The Bank of International Settlements as a think tank for financial policy-making. *Policy & Society, 37*(2), 170–187.

Whyte, J. (2013). *Quack policy*. London: IEA.

Williams, B. (2015). *The evolution of Conservative Party social policy*. London: Palgrave Macmillan.

Williams, R. (2012). The rolling apocalypse of contemporary history. In J. Caraça, G. Cardoso, & M. Castells (Eds.), *Aftermath: The cultures of economic crisis* (pp. 17–43). Oxford: Oxford University Press.

Wolf, M. (2014). *The shifts and the shocks: What we've learned—And have still to learn—From the financial crisis*. London: Penguin.

WonkComms. (2014). *Taking the pulse: The role for evidence in the election debate*. Accessed 30 October 2015. https://www.youtube.com/watch?v=wI2Sv8mD-3k.

Worstall, T. (2010). *Chasing rainbows: How the green agenda defeats its aims*. London: Stacey International.

Wren-Lewis, S. (2018). *The lies we were told: Politics, economics, austerity and Brexit*. Bristol: Bristol University Press.

Yin, R. (1994). *Case study research: Design and methods*. Thousand Oaks: Sage.

Index

M. González Hernando, *British Think Tanks After the 2008 Global Financial Crisis*, Palgrave Studies in Science, Knowledge and Policy, https://doi.org/10.1007/978-3-030-20370-2

Y

The manufacturer's authorised representative in the EU is Springer Nature Customer Service Centre GmbH, Europaplatz 3, 69115 Heidelberg, Germany. If you have any concerns regarding our products, please contact ProductSafety@springernature.com

Printed and bound by CPI Group (UK) Ltd, Croydon, CR0 4YY

22/09/2026

02221067-0011